Sustainable Smart Cities - A Vision for Tomorrow

Edited by Amjad Almusaed and Asaad Almssad

Published in London, United Kingdom

Sustainable Smart Cities - A Vision for Tomorrow
http://dx.doi.org/10.5772/intechopen.100727
Edited by Amjad Almusaed and Asaad Almssad

Contributors
Doghmane Mohamed Zinelabidine, Eladj Said, Fachrul Kurniawan, Mochamad Hariadi, Supeno Mardi Susiki Nugroho, Prita Dewi Basoeki, Toshio Uchiyama, Tsukasa Hokimoto, Jong Woo Choi, Jae Eun You, Menachem Domb, Kamal Shahid, Muhammad Hassan, Sadaf Mukhtar, Ali Husnain, Chang-Wei Chai, Yu-Heng Huang, Tseng-Wei Chao, Margherita Pillan, Lasse Berntzen, Qian Meng, Adipandang Yudono, Sapto Wibowo, Christia Meidiana, Surjono Surjono, Irnia Nurika, Erryana Martati, Yan Akhbar Pamungkas, Daniel Adu, Ransford O. Darko, Boamah Kofi Baah, Agnes Abbey, Laurentiu Constantin Lipan, Sorin Dimitriu, Amjad Almusaed, Asaad Almssad

First published in London, United Kingdom, 2023 by IntechOpen
IntechOpen is the global imprint of INTECHOPEN LIMITED, registered in England and Wales, registration number: 11086078, 5 Princes Gate Court, London, SW7 2QJ, United Kingdom

British Library Cataloguing-in-Publication Data
A catalogue record for this book is available from the British Library

Additional hard and PDF copies can be obtained from orders@intechopen.com

Sustainable Smart Cities - A Vision for Tomorrow
Edited by Amjad Almusaed and Asaad Almssad
p. cm.
Print ISBN 978-1-80356-764-8
Online ISBN 978-1-80356-765-5
eBook (PDF) ISBN 978-1-80356-766-2

For EU product safety concerns:
IN TECH d.o.o., Prolaz Marije Krucifikse Kozulić 3, 51000 Rijeka, Croatia,
info@intechopen.com or visit our website at intechopen.com.

Meet the editors

Prof. Amjad Almusaed has a Ph.D. in Architecture (Environmental Design) from Ion Mincu University, Bucharest, Romania. He completed postdoctoral research on sustainable and bioclimatic houses at the School of Architecture, Aarhus, Denmark in 2004. His academic focus is on the role of sustainability in building design and city planning. He has done extensive study, research, and technical surveying in these fields. He is an active member of numerous international architectural organizations and has edited several foreign books. His bibliography includes around 185 worldwide academic publications (articles, studies, books, and book chapters) written in various languages.

Associate Professor Asaad Almssad has worked in industry, academia, and research for more than 30 years. He has held positions at Umea University, Sweden; Karlstad University, Sweden; and a variety of other European and non-European institutions. His studies concentrate on building materials, building structures, environmentally responsible construction, and the energy efficiency of building systems. He has more than sixty academic papers and numerous books to his credit Currently, he is employed as a docent at Karlstad University, Sweden.

Contents

Preface

Today, urban areas are inhabited by more than half of the world's population. More than 70% of worldwide carbon emissions and 60%–80% of energy usage take place in metropolitan areas, according to the Organisation for Economic Co-operation and Development (OECD). Governments and municipalities may create more intelligent, sustainable cities using ICTs and other technology. To address current and future generations' economic, social, environmental, and cultural demands, a municipality must be innovative and use ICT to boost living standards, city performance and services, and competitiveness. Although there is not yet a city where every system and function are linked together, many are well on the way to becoming intelligent, sustainable cities. They utilize technology to reduce energy consumption and waste, increase traffic safety, monitor air quality, report crimes in progress to police, and upgrade water and sewage systems, to name a few examples. Many ICT-based applications and services rely on a stable, secure, dependable, and interoperable telecommunications infrastructure, which is essential for smart, sustainable cities.

The IoT is a rapidly growing network of computing devices with built-in sensors and software that can connect and share data, allowing billions of devices and objects with smart sensors to connect, collect information in real-time, and wirelessly communicate this data to a centralized control system. These networked control centers oversee traffic flow, reduce energy use, and enhance a wide range of municipal services. Computational analysis of massive data sets is made possible by AI, revealing patterns that may be utilized to inform and improve municipal decision-making. For example, electricity networks called "smart grids" employ digital communication technology to monitor and react to fluctuations in local electricity usage, allowing for more efficient use of electricity and a reduced environmental impact. Users have more say over their energy consumption with the help of IoT devices like smart meters and sensors that report back to energy providers on consumer energy habits. Thoughtful city planning, predictive maintenance of city services, real-time monitoring, decision-making, and optimization of emerging technologies across different industries like AI, the IoT, and others are all made easier with the help of digital twins, which employ virtual and augmented reality to do the same. When compared to the current 3G and 4G networks, which have issues supporting the variety of services needed for sustainable smart city applications, the rollout of 5G, the fifth generation of mobile technology in many countries, has the potential to connect more devices to the Internet, the possibility to transfer data faster, and the potential to process large amounts of data with minimal latency.

While no "smart" or "sustainable" communities are wholly wired for all municipal systems and services, many are well on their way. Some urban areas, for instance, use ICT to better manage their energy resources, garbage, housing stock, medical facilities, transportation networks, air quality, alert systems for criminality on the streets, and water and sewage systems. Improving the quality of life for people living in rural areas and helping to accomplish the United Nations' Sustainable Development Goals (SDGs) are both possible outcomes of rural community development that are both smart and sustainable.

The International Telecommunication Union (ITU) funds technical research and encourages continuing collaboration between governments, industry firms, and other stakeholders to boost the dependability, security, and interoperability of urban ICT infrastructure. The ITU is promoting information and communication technologies to improve the efficiency of services and living standards in metropolitan areas. Improving the quality of life for rural residents and contributing to the SDGs can be aided by, for instance, innovative construction and sustainable rural communities. Among the seventeen SDGs is SDG 11, which focuses on creating sustainable cities and neighborhoods. ICT can speed up the process of achieving this goal as well. SDG 11 emphasizes the need to develop environmentally friendly urban environments. The concept of a "Smart City" and the theories behind it have come a long way in recent years. This is in large part because so many branches of government are investigating the potential of "smart city" technologies. The United Nations, the European Union, and other federal ministries have also tackled the issue of urbanization to improve a city's quality of life via ICT and to satisfy present and future generations' economic, social, environmental, and cultural demands.

The book discusses the structure and operation of a sustainable smart city, a model for the next generation of urban areas. To maximize resource utilization efficiency, enhance urban management and services, and better the lives of its residents, "smart cities" implement various information technologies and novel ideas to integrate the system and services into the town. All facets of city life make use of the latest generation of information technology. It accomplishes the refined and dynamic management of urbanization, which helps reduce "big city illness," improve the quality of urbanization, increase the efficiency of urban governance, and better the lives of residents. These urban forms invest in human and social capital as well as transportation and information communication infrastructure, practice participatory management of these and natural resources for scientific management, all of which contribute to the city's ability to grow economically and socially sustainable and to provide its residents with a high quality of life.

This book presents the modern concept of a "sustainable smart city" based on smart management, smart human environment, and smart technologies of urban communications. It explores the vision, planning, and action strategy of a smart city. In addition, it presents practical examples of smart city building, detailed explanations of the elements influencing construction decisions, and the methods used to narrow those choices. Readers involved in building intelligent cities as well as government officials, company managers, scientific researchers, college lecturers, and students will find this book a useful resource. It includes thirteen chapters, each presenting a particular component, system, or technology essential to developing a next-generation smart city that can function without compromising its environmental or economic sustainability. Writing a book takes time and work, just like developing a construction concept does, and requires the cooperation of many individuals over a prolonged period. We must start by giving each other credit for our contributions to our joint venture. As such, I offer my deepest gratitude to every one of the contributing authors for their hard work. In addition, I would like to thank Author Service Manager Ms. Martina Scerbe at IntechOpen for her assistance and professionalism during the publication process of this book.

Amjad Almusaed
Department of Construction Engineering and Lighting Science,
Jönköping University,
Jönköping, Sweden

Asaad Almssad
Faculty of Health, Natural Sciences, and Technology,
Department of Engineering and Chemical Sciences Construction Technology,
Karlstad University,
Karlstad, Sweden

Section 1

Smart Cities Models

Chapter 1

Blending Human Ware with Software and Hardware in the Design of Smart Cities

Amjad Almusaed and Asaad Almssad

Abstract

Sustainable innovation entails realizing society-oriented value creation in an environment-friendly manner. A smart city can be viewed as a holistic paradigm that avails of state-of-the-art information and communication technologies (ICTs, in other words) to advance the so-called "Internet of Things." This aids the management of urban processes and improves the quality of life for the citizens. Smart cities are bound to keep getting "smarter" as the ICTs keep developing. While the technological factor represented by the IoT, augmented and virtual reality, artificial intelligence, urban digital twinning, cloud computing, and mobile Internet is a driving factor unarguably, innovation in urban ecology is a vital socio-economic factor that will spur the transformation of urban areas in the world to smart cities. In this chapter, the authors answer the "what," how, and "who," so to say, of the paradigm—smart cities—with real-life examples and a case study. They emphasize the importance of human ware and remind readers that technology—the all-encompassing Internet of Things with its infantry of cameras, sensors, and electronic devices—though powerful, is a humble servant in the service of the inhabitants of a smart city.

Keywords: artificial intelligence, cloud computing, internet of things, smart cities, urban digital technology, urban spaces

1. Introduction to the paradigm: smart cities

Functionally, one may identify residential, industrial, and commercial areas within a city. The city government and the commercial facilities are usually centralized in the so-called "city Centre" or "central business district," while the residential areas (inner city) and industrial complexes are distributed over the surrounding land area [1]. According to the United Nations (2016), by 2030, 60% of the global population will be urbanized [2]. The paradigm "smart city" was conceived in 2008, when IBM created a plan for the Smart Planet project to build new cities that could support a burgeoning human population, while also enhancing the quality of life for their inhabitants. Leading IT corporations jumped on the bandwagon and the concept entrenched itself. Many countries—Singapore, the United Arab Emirates, and South Korea to name but three, have invested a lot in their smart city initiatives. Songdo (South Korea) can be looked upon as the very first turnkey smart city. Cities, in general, are hubs of creativity and innovation, which stand them in good stead to adapt to/counter/minimize/solve problems/challenges related to rapid urbanization, including issues with social cohesion, the demand for natural resources, the effects of climate change, and rising demand for city services such as transportation, health, housing, and social care [3, 4]. The development and integration of ICTs remove obstacles to the exchange of knowledge and information, and restrictions on innovation while encouraging the dissolution of barriers between different social organizations and activities. The transition from the production paradigm to the service paradigm positively impacts the "industrial form," "city-administration form," and the "urban form" in general [5]. The idea of a "smart city"—from a technocratic perspective—is to manage the inanimate assets in the urban setting to serve the animate entities (human inhabitants) by integrating various ICTs (information and communication technologies) and IoT solutions. The assets include local information systems departments, schools, libraries, transportation, hospitals, power plants, water and waste management utilities, law enforcement agencies, and other public services [6–8]. By utilizing urban informatics technology to improve the efficiency of service provision, and cater to the ever-changing demands of the inhabitants, a smart city strives to make living healthier, safer, more prosperous, comfortable, and enriching for its citizens, by gathering data continuously and promptly addressing any

Figure 1.
The six main smart city elements [10, 11].

issues of inefficiency that may crop up [9, 10]. The Center for Regional Science at the Vienna University of Technology has identified six key characteristics of a smart city (encompassing all the pillars of sustainable development), which provides a useful foundation for choosing dimensions while considering a particular city's resources and long-term objectives (refer **Figure 1**) [10, 11].

- Smart environment
- Smart mobility
- Smart living
- Smart people
- Smart government
- Smart economics

Needless to state, to use a metaphor, smart government is akin to the lubricant which keeps the intermeshing gears of environment, mobility, living, people, and economics rotating in tandem. Improvements of [11] and changes in the digital infrastructure [10] are under the purview of "smart government."

What is a smart city?

The "what" if "smart cities" can be comprehended well, by resorting to published literature. What follows is a bulleted list carefully compiled from relevant literature sources.

- The definition of a smart, sustainable city is "an innovative city that uses information and communication technologies (ICTs) and other means to improve quality of life, the efficiency of urban operation and services, and competitiveness, while also guaranteeing that it meets the needs of current and future generations concerning economic, social, and environmental aspects" [12].
- A "smart" city, or Smart Municipal, on the other hand, is a man-made interconnected system of information and communication technologies with IoT, or the internet of things, which streamlines the administration of internal city activities and improves the quality of life for citizens.
- A city that aspires to become a smart, sustainable city must, in theory, improve its attractiveness, sustainability, and inclusivity, for inhabitants (permanent and temporary) [13, 14].
- According to the Smart City Council, "A smart city incorporates digital technology in all the functions of the
- A city can be defined as "smart" when sustainable economic development with intelligent management with investments in human and social capital and traditional (transport) and modern (ICT) communication infrastructure and promote a higher quality of life. Natural resources are used through participatory action and engagement [15].

- A smart city is "the successful integration of physical, digital, and human systems in a built environment to ensure a sustainable, prosperous, and inclusive future for residents," according to the British Standards Institution (BSI) [16].

- There are eight key aspects that define a smart city, according to Frost and Sullivan (2014): smart governance, smart energy, smart building, smart mobility, smart infrastructure, smart technology, smart healthcare, and smart citizens [16].

- A smart city brings together technology, government, and society to enable the following features: a smart city, a smart economy, smart mobility, a smart environment, smart people, smart-living smart governance [17].

- The concept is not static; there is no absolute definition of a smart city, no end point, but a process, or series of steps, by which cities become more "livable," more resilient and, therefore, able to respond more quickly to new challenges [14].

- Smart cities represent a unique specific entity—something that can be taken as a whole and launched as a comprehensive itemized list. All hardware must be linked to the internet of things, all software must be connected, and every component must be instantaneously updated and synchronized for a city to be deemed smart [18].

- This group of cities may be compared to a well-functioning biological creature that promotes social welfare. Digital smart city technologies make life better for inhabitants on all levels. For example, the time wasted in slow-moving traffic is almost eliminated, parking spots for private vehicles can be easily located, and inhabitants feel more secure and safer. In a nutshell, smart technologies will make life more convenient and comfortable, for the urbanites of the world (55% in 2018 and expected to be 60% by 2030) [19, 20].

- The main idea of a smart city is the prevalence of ICTs and the IoT to gather data on energy and water usage, vehicular traffic, air pollution, and other urban "variables" in order to plan, make decisions, change and evolve into a more sustainable urban setting, utilizing resources—be they water, energy, food or for that matter, urban space, optimally [19, 20]. However, authors of Ref. [21] point to the shorter lifetimes of the smart network devices (necessitating frequent replacements) and the high energy consumption associated with the data servers for instance.

2. Wise use of the limited resource: urban space

The need of the century is an agglomeration of urban areas generating sustainable economic development and contributing to social welfare (enhancement of quality of life, in other words), by availing of the six key "smart" characteristics which have been referred to earlier, and thus adapting to or surmounting the sustainability-related challenges of the century [22, 23]. The increase in urban land usage is sometimes referred to as urbanization. The traditional definition of urbanization considers "land-use change" from scattered "exploitation" of the resource to more compact land-use practices [24]. It is an assemblage of architectural and engineering artifices that enable the city's permanent and transient residents to perform their daily functions. Mythologically, the city was looked upon as an analogue of the model of the world—the heavenly world in earthly manifestation, in other words. *"The city is the connection of heaven and earth, and we live in it,"* is one of the many inscriptions on tablets from the Sumerian civilization [25]. The city spaces represent a system that presupposes the presence of material/s (what the studied phenomenon or phenomena consist of/s of). The city spaces can be guided by anthropocentric logic—in other words, place a person at the heart of the urban planning process. The materials in the space will then obviously have the "fingerprints" of the inhabitants, so to say. These "fingerprints" are the needs and values of the human entities of the system [26]. In the past, cities were created to provide safety and defense against outside threats for their inhabitants. Following this, the inhabitants began to group together and become "fellow citizens" to support trade. The purpose of "public spaces" in urban areas is the facilitation of social interaction and communication. These spaces need to express empathy for human needs so that people of any socioeconomic class may feel comfortable when they spend their leisure time there. Using a thermodynamic metaphor, high-quality and lively public spaces provide both sensible and latent benefits to the populace [27, 28]. They serve as incubators for urban development and must be designed/created in keeping with the aspirations of its "users." Practice shows that the most popular urban spaces are multifunctional, providing visitors with options for several leisure activities, and by doing so, attracting people from different walks of life. In today's globalized human society characterized by fluidity and diversity, the approach to urban spatial and structural organization has changed considerably and is oriented toward unification, optimization, and digitalization [29]. Society needs new views and strategies in the context of urban evolution to understand and corroborate the requisites for a "human-friendly, safe and comfortable urban environment," and the modus operandi to get there [30, 31]. Many researchers are actively considering the future of cities in terms of the digitalization of society and the introduction of IT technologies, believing that these processes can influence the creation of comfortable and conducive social conditions. In the process, less attention is paid to the interactions among the denizens of the city. Its significance in the formation and development of individual identities is overlooked. The influence of the environment—both natural and anthropogenic—on society (the society-environment nexus, in other words) cannot be ignored or denied. Cities are centers of intellectual activity, commerce, culture, science, productive labor, social development, and much more [32]. Nevertheless, they are also plagued by a host of challenges, triggered by population growth—overcrowding, lack of housing, lack of funds to provide basic services to the population, and degradation of infrastructure [33]. If these challenges are not addressed pronto and tackled head-on, there is a clear risk of rising discontent, escalating political and racial conflicts, and a spike in the crime rate [34]. What lies ahead for urban planners and city administrators is a gargantuan task Clever, out-of-the-box approaches may ease the way forward a little—utilization of the resources and ideas of the neighborhood to design areas that seamlessly and naturally blend into the urban fabric. The motivator here should be the fostering of a sense of community, *via* creative uses of urban space—like for instance, converting an ancient town square for new purposes, or by rebuilding a park on a site that houses an abandoned factory.

2.1 Modern cities and the "smart city" model

Many cities have implemented "smart city" (hereafter written without quotation marks) policies. Smart city conceptualizations of cooperation place a strong emphasis on a strong inter-stakeholder rapport, which is indispensable for effective collaboration towards common goals. Apart from inter-stakeholder liaison (involving the government, inhabitants, industries, banks, media, academic institutions, commercial entities, etc.), there is also a need for inter-departmental collaboration at the governmental level [35]. "Smart" entails "transparency" and thereby the availability of data and information across open-access networks to inhabitants of the city. Globally, smart cities seek leaders with foresight, who are effective team players. Over the last decade, the concept of "smart cities" has gained significant popularity in policy and research circles. However, as it evolves, it needs to adopt a more citizen-centric approach, instead of being a slave to technology [23].

2.2 Digital city infrastructure

Digital city infrastructure comprises the fundamental information technologies, organizational structures, and associated services and facilities required for a business or industry to operate in a smart city. A given urban infrastructure may be linked to other cities and countries, forming in the process, what could be labeled as regional, national, or global infrastructures. If specific to the corporate world, one could speak of industrial or corporate digital infrastructures [36]. Such infrastructures are complex systems consisting of many subsystems, networked computers, controllers, sensors, and devices, which amass and crunch data, and transmit processed data, alternately called "information" [37]. The digital infrastructure rides over and thereby monitors the physical infrastructure, which includes roads, bridges, parks and buildings, security and safety systems, HVAC systems, water and sanitation networks, power supply systems, *inter alia*.. Digital infrastructure is critical at facilities where a range of services are offered by a host of service providers [38]. Institutions are progressively compelled to reassess their current capabilities, structures, and cultures to uncover possibilities to incorporate state-of-the-art technologies, in the process of overhauling existing work models [39]. Coordinating the multiple tasks happening concurrently within institutions is indisputably a complex, time-consuming, energy-intensive task. Consider this as an example—simultaneous operation of both heating and air conditioning systems on the premises of a firm. The implementation of a digital (city, corporate, or industrial) infrastructure eliminates the complexity of operating multiple systems simultaneously and results in some cost reduction too in the process. Avoiding redundancy by using one single network for the transmission of video, voice, and data is cost-effective [40]. The digital infrastructure available today needs to evolve to meet the ever-changing needs of urban residents, related to the nine components of the digital/physical infrastructure shown in **Figure 2**.

Future smart cities will need to manage almost in real-time, optimize their resource usage, boost mobility, lower noise, and pollution levels, provide easy access to online services, have smart buildings that draw visitors, improve the safety and security of its citizens, and create new economic opportunities. This will necessitate harnessing ICTs, and monitoring and measuring to be able to manage [41]. While data networks are mandatory, data privacy issues cannot be swept under the carpet [42]. The definition of a smart, sustainable city is "an innovative city that uses information and communication technologies (ICTs) and other means to improve quality of life, the efficiency of urban operation and services, and competitiveness, while also guaranteeing that it meets the needs of current and future generations concerning economic, social, and environmental aspects" [43]. A "smart" city, or Smart Municipal, on the other hand, is a man-made interconnected system of information and communication technologies with IoT, or the internet of things, which streamlines the administration of internal city activities and improves the quality of life for citizens. A city that aspires to become

Figure 2.
Smart city infrastructure components.

a smart, sustainable city must, in theory, improve its attractiveness, sustainability, and inclusivity, for inhabitants (permanent and temporary) [44, 45]. Real-time monitoring generates data collected from citizens (households) and other items of infrastructure in the city (refer **Figure 3**). These data are stored on computer systems in data centers, which can be a group of buildings housing mobile systems and associated components [46]. The data security issues referred to earlier, are very crucial for these data centers; likewise, the energy consumption by the servers which hold the unimaginably huge volumes of data is also a matter of concern [44], especially in smart cities, which have a substantial portion of their electricity being sourced from fossil-fuel-powered thermal power plants. However, the data centers can be equipped with their own renewable power production units for captive consumption, if possible and feasible.

Our understanding of local city dynamics is evolving thanks to smart cities. To design public policies oriented toward improving the quality of life for the citizens, thorough, holistic planning is necessary. All stakeholders involved, all types of resources demanded, and all items of infrastructure need to be factored in, to harness the synergies, and minimize the tradeoffs/conflicts [47]. Personalized smart cards owned by the inhabitants and used at various "points-of-sale," parking lots, public transportation systems, etc., are vital components of the "smart networks" in smart cities. The use of these smart cards enables the city planners to analyze the behavioral patterns of the inhabitants and utilize this knowledge as the basis for decision-making focused on modifications to, and improvements of the city infrastructures [48]. Smart city programs are being implemented (note that this is of a dynamic nature, and is continuous) at the time of writing, in Amsterdam, Barcelona, Madrid, Stockholm, Chicago, Beijing, Glasgow, Dublin, and various cities in India; and there is thus a gradual proliferation of smart meters, smart grids, smart residences, and smart buildings [49]. **Figure 4** illustrates the four characteristics of smart buildings/residences [49].

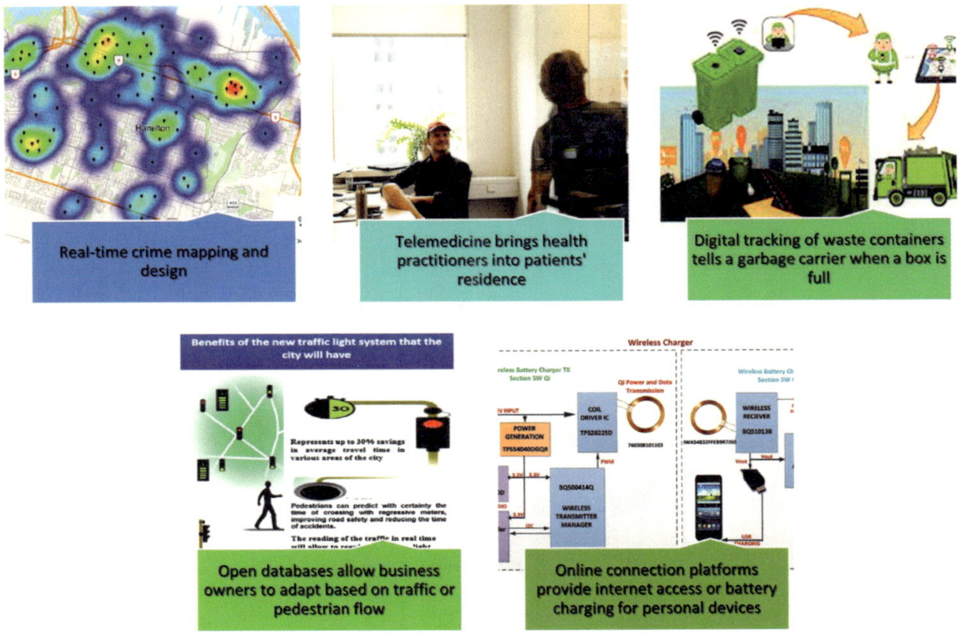

Figure 3.
Practical application in different development areas [46, 47].

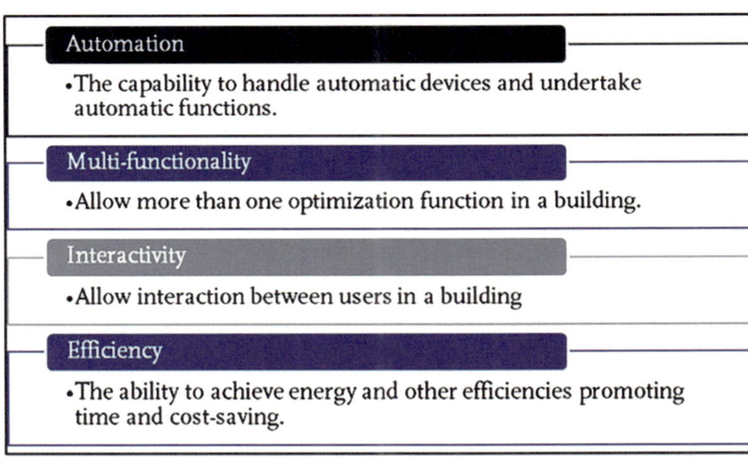

Figure 4.
The four features of smart buildings [49].

3. Smart facility management

3.1 Internet of things (IoT) in smart city

The term "Internet of Things" abbreviated as IoT was coined way back in 1999 by entrepreneur Kevin Ashton, who co-founded Auto-ID Labs (an independent laboratory network and research group in networked radio-frequency identification devices and new sensor technologies) at the Massachusetts Institute of Technology. IoT, which has now come to stay, is often promoted as the next significant advancement in massively dispersed information, allowing any physical device to automatically join online and be searched for by anybody in the world [50]. The Internet revolution has occurred in four different phases—the first three focused on specific devices [51]. Through the IoT (the fourth phase, or wave), these devices that have become part and parcel of the anthroposphere, are all connected to each other directly or indirectly—usually securely—to enable centralized monitoring and control, quick responses to emergency situations, and proactive/reactive strategizing. Indeed, nothing is perfect. There are loopholes and risks and data privacy, as also referred to earlier, is a concern that needs to be addressed with vigil [45]. Intelligent computers which can comprehend, learn, and perform human-like activities and be trained to modify tasks over time to increase accuracy and efficacy are a part of what is termed as artificial intelligence or AI [49]. AI, thus, is a part of the IoT. One application is the control and optimization of the amount of energy used for lighting and heating. Another example is the concept of a "smart factory," in which automated guided vehicles (AGVs) monitor industrial equipment, look for problem areas, and then rearrange themselves to forestall and obviate breakdowns [52].

Traditionally, connectivity relied primarily on Wi-Fi, but today, 5G and other types of networking platforms are becoming more efficient at managing large datasets and providing speed and reliability [53]. The utilization of the data, not the data itself, is the primary goal of data collection. IoT devices gather and send data, which must be very carefully examined to make wise decisions (see **Figure 5**). IoT is slated to grow thanks to developments in AI-supported machine learning and superior analytics [53, 54].

3.2 Artificial intelligence in smart cities

Humankind strives to create better-living conditions in cities, and technological advances have brought about rapid transformations over time [55]. A modern smart city, while developing sustainably, must respect the planetary boundaries and ensure the socio-economic welfare of not just the existing population but also the generations to follow. Decision makers must continuously be aware of the connections, synergies, and trade-offs among these pillars to uphold and advance the principles of this paradigm in the interest of human development and ensure responsible human behavior and actions at the global, national, community, and individual levels [56].

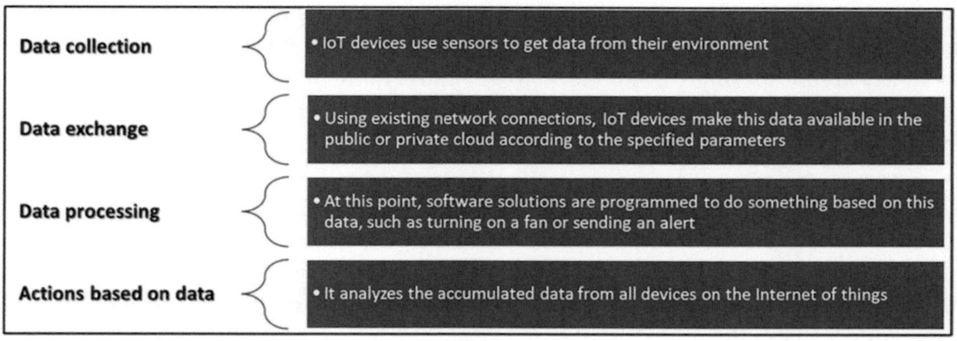

Figure 5.
Smart cities working process.

Artificial intelligence (AI) is a powerful tool in the "toolkit" of urban planners. Smart algorithms are currently helping many organizations in the private and public sectors to improve their operating efficiencies [57]. In the public sector, for instance, city traffic management can be improved significantly by availing of an adaptive AI-based traffic management system [58]. Strategic placement of sensors closes to street lamps/ bulbs in public spaces facilitates data gathering. According to a new study, a new breed of intelligent lampposts that can monitor body temperature and recognize crowding may be able to stem the spread of COVID-19 and restore communities [59, 60]. They could also consist of 5G Wi-Fi hotspots, air quality sensors, flood monitors, digital signs, and video surveillance systems. For instance, the municipal council of Barcelona has created a camera-based system mounted on lampposts in the Las Ramblas region to aid crowd management and monitor public health on the beaches [61]. The municipality used scanning devices to get the images and some AI to analyze them to figure out how much of the beach is free. Smart solutions do not have to be universal, perfect, and extremely expensive. Any small improvement using AI, albeit far from perfect, can bring in a lot of value. Organizations must comprehend the value AI technologies can bring to their operations. However, current AI research is more concerned with understanding how AI is adopted technologically than finding its use-related organizational issues [62]. China, for instance, has seen tremendous economic expansion and hyper-rapid urbanization over the last three decades, aided by ICTs, cloud computing, and the IoT [63]. Alibaba's ET City Brain 2.0 is an AI-based traffic management system first adopted by the Hangzhou city administration to report violations of traffic rules in real-time and provide unhindered passage for emergency vehicles such as fire engines, in the event of emergencies [64, 65]. It goes without saying that machines, unlike humans, do not get tired while performing repetitive tasks such as checking the identity of passengers at airports. Repetitive manual work is often error-prone owing to the resulting tedium [66, 67].

AI has sparked controversies around the world, and many of them have not been resolved. Inhabitants are averse to being monitored and label the presence of video cameras collecting data, as an infringement of their right to privacy. However, responsible city administrations must impress upon people the indispensability of AI-based technologies to support intelligent decision-making to ensure greater safety, security, and comfort [68]. The data gathered by IoT applications are typically unstructured. AI-based models extract relevant data from huge volumes of diverse datasets and facilitate focused learning therefrom [69]. As artificial intelligence has grown and its demand for data has expanded, the number of IoT devices has substantially increased. It is common to undervalue the promise that applications of artificial intelligence offer for "smart cities.". At the time of writing, the taxonomy of smart city indicators is under discussion. Additionally, since the concepts of sustainability and resilience are increasingly understood to be linked to the idea of the smart city, more clarification is needed regarding how various assessment frameworks or indicator sets are aligned with sustainability and resilience dimensions and characteristics [23].

3.3 Augmented and virtual reality (AR/VR) in smart cities

ICTs are not without their challenges; they have spawned cybersecurity concerns. Although augmented reality (AR) and cyber-security technologies have been around for a while, of late, they have experienced exponential growth [70, 71]. Augmented reality will be an integral part of the digital infrastructure of smart cities in the years to come [72].

Virtual reality or VR has come a long way since 1961 when it was first introduced for military applications in the USA. The deployment of AR and VR in urban environments offers several benefits, including easy navigation and a good knowledge of the specifics of urban life. It can increase public participation in urban decision-making and contribute to a collaborative urban design process [73, 74]. A "smart city" does not just imply the accessibility of municipal services online, but also entails a deeper integration of such services and automated systems, facilitating proactive asset management and the wise use of urban space [74] (and customer relationship management, where the inhabitants are the "customers" of the city administration). **Figure 6** illustrates the six areas in which AR and VR can be effectively deployed in smart cities.

It is beneficial to realize that the primary characteristic of an urban environment is the concentration of social activity. Therefore, it is necessary to analyze statistics on the environment, commercial activity, and the use of public spaces by the citizens, in addition to traffic data from streets, electrical networks, and water supplies. The social, economic, and biophysical surroundings have an impact on how people act and interact, and thereby on health and quality of life [75]. Visualization helps city authorities to understand the prevailing situation better and make improvement decisions based on such understanding. All these concepts and elements are presented as layers of the urban geographic information system (GIS). The connection with urban GIS is the most understandable and well-developed use of VR and AR in smart city technologies. At the same time, Web virtual reality (WebVR), IoT, and three-dimensional (3-DGIS) geographical information system (3-DGIS) with peer-to-peer (P2P) networks are some of the most recent integrated IT tools. These are useful while handling spatial "big data" (remote sensing data) [76]. Much has been written about the potential of smart urbanism to bring about varied and permanent types of progress, including favorable energy efficiency improvements and a heightened sense of environment friendliness [77].

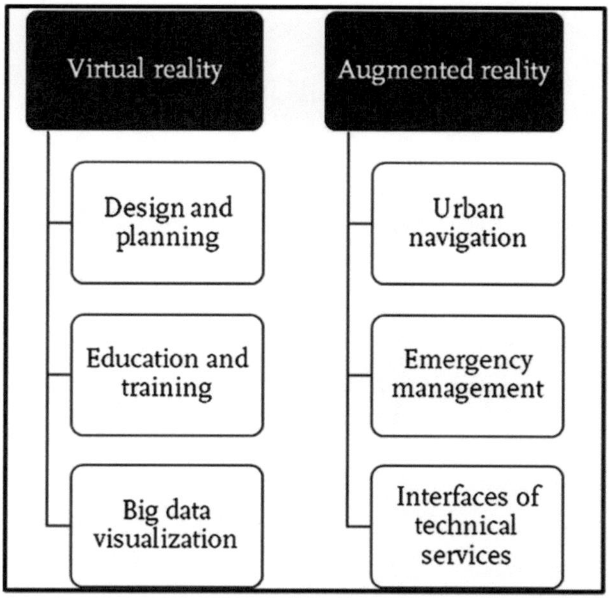

Figure 6.
The six critical areas for the use of virtual reality/augmented reality in a smart city [73, 74].

3.4 The urban digital twin technology

The urban digital twin (UDT) technology combines 3D city models with dynamic data from sensors and GIS technologies, and contributes to a much-better understanding of cities [78]. This strategy is based on the idea of the possibility that a static structure with dynamic features will arise. Although this technique has industrial and technical roots, NASA employed it for the first time in the 1960s to physically duplicate systems on Earth to match those in space. The digital twin is now within reach as production and manufacturing become increasingly digital, and the IoT becomes all-pervading. Digital twins are created to interact with their environment in several ways to replicate complex structures and processes for which it is challenging to predict effects throughout the product's existence.

This link between the actual and virtual worlds occurs almost instantly, and more precise forecasts can be made well in advance to enable adaptation or preventative management [79]. Due to its low cost, rapid analysis, minimal risk, and potential for substantial insight, the simulation of manufacturing systems is a potent tool for the analysis of systems, and an understanding of the roles of and the interactions among the components thereof. A software equivalent of a physical item mimics a real thing's inner workings, technical details, and behavior [80]. The digital twin's use of sensor data from an actual device working in parallel to set input actions on it is a critical component of data-driven decision-making, monitoring of complex systems, product validation and simulation, and object lifecycle management. Both offline and online modes of working are possible. Additionally, information from the digital twin's virtual sensors and the actual device's sensors may be compared to find abnormalities and the causes thereof (**Figure 7**) [82].

Although there has been a proliferation of publications related to UDT (or DT, rather), any investment in such technology needs to be made, after a thorough understanding of the requirements, purposes to be served, and the benefits and limitations of the technology [83]. The academic sector, businesses, and the public transportation

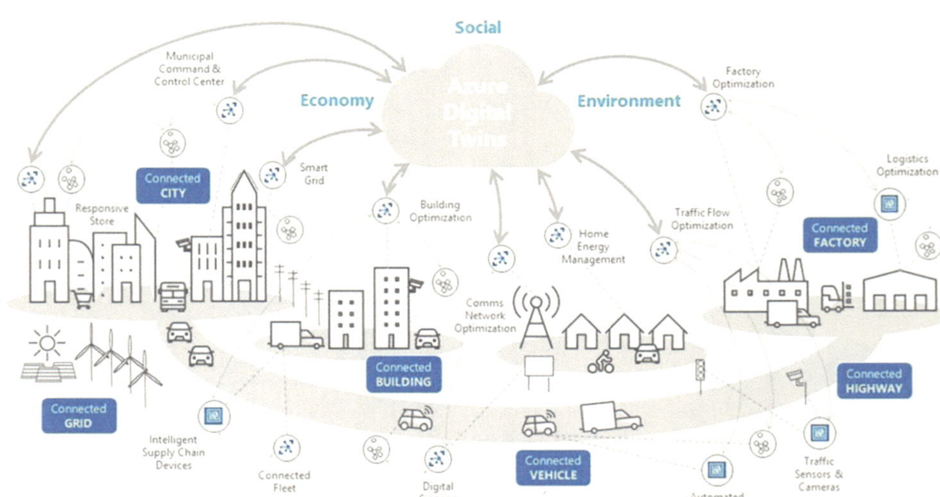

Figure 7.
Digital twins, environments of all types [81].

sector are particularly interested in DT, as it holds promise to help users surmount the challenges posed by tighter profit margins, and stringent quality demands imposed by regulators and consumers, and productivity improvements expected by investors [84]. Creating smart cities is a complicated process involving a web of interactions among municipal departments, external stakeholders, and a wide range of service providers. City planners may create smart, sustainable, safe, and liveable smart cities with the use of two technological practices: urban information modeling (CIM) and city digital twins (UDT) [85].

3.5 Urban drones in smart cities

Drones are vital components in smart city networks, supporting a plethora of functions (see **Figure 8**) ranging from delivery of packages to policing to traffic monitoring to firefighting to rescue operations during natural calamities [88]. However, due to the absence of established algorithms for using urban drones in both standard and special circumstances, it is desirable to assess the drone-related experience and identify the most valuable strategies for future best practices [89]. When traffic and street cameras do not serve the purpose or are absent, drones can step in as replacements. The quadcopter can provide real-time video surveillance and help to avert natural catastrophes, investigate traffic accidents, photograph scenes of crime and gather evidence, and detect faults in infrastructures. The features of urban drones differ based on the platform and the intended application, and thereby a classification must take into consideration a wide range of factors [90]. Most of the drones studied, utilized single-rotor, rotary-wing drones with cameras serving as aerial sensors [91]. Perhaps, using drones to deliver packages can reduce traffic congestion (as drones would be replacing road vehicles which would otherwise be deployed for the courier

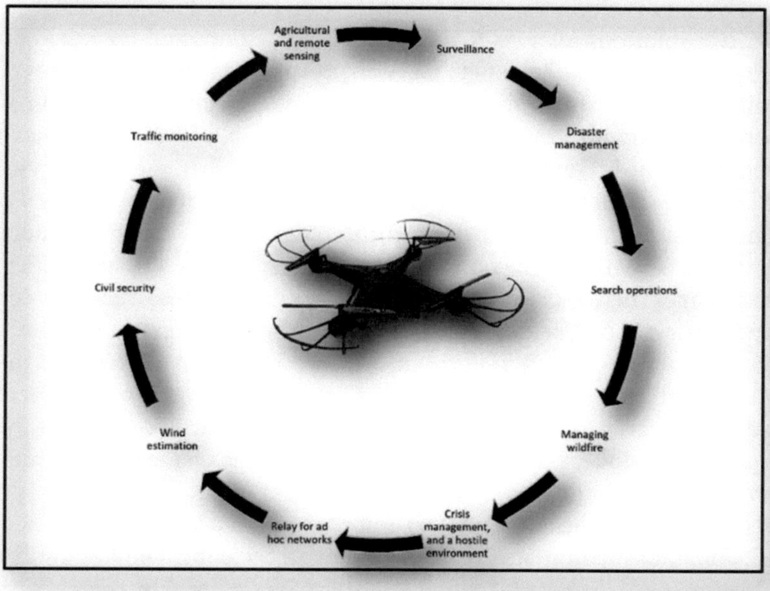

Figure 8.
The role of drones in a smart city [86, 87].

service), curtail air pollution [86], save time [87], and venture into areas that would be a wee bit difficult for rescue workers to directly get to [92]. Cumulatively, over time, the benefits would be conspicuous.

4. Application of the IoT

4.1 Cloud computing and peripheral computing

Applications connected to the IoT have become interesting areas of research for engineers and academics. This testifies to the size and significance of data-related issues that must be resolved in modern commercial enterprises, particularly in cloud computing. Just like network connections, the expansion of cloud computing is strongly tied to the growth of the IoT. Due to its ability to provide processing power and on-demand storage for big data, IoT cloud services have allowed devices to capture, store, and transmit larger and more complex sets of data [93, 94]. Additionally, while retaining the security of a closed system, private cloud solutions have allowed businesses to handle significant volumes of different types of IoT data. Traditional urban management techniques and technologies can no longer keep up with cities' demands and diversified development needs—road traffic management, medical and health services, and public service facilities [94]. Just as network connections have been closely tied to the growth of the IoT, cloud computing too is. Because IoT cloud services can provide processing power and on-demand storage for big data, applications can be hosted in the cloud. A cloud computing service provider makes public cloud services available, allowing third parties to update the IoT environment and incorporate data into IoT-enabled electrical products [95]. Urban architectural standards have evolved over time and are often in a state of flux, which is due to the rapid alteration in the influential factors [96–98]. The sustainable development goals (SDGs) of the 2030 Agenda for Sustainable Development of the United Nations include, among other things, making cities more sustainable and resilient. ICT solutions relevant to different urban systems and domains come into play here [99]. Most urban stakeholders (service providers as well as service receivers) see cloud computing as an emerging and attractive strategy for healthcare. It has the unique capability to provide limitless capacity and power of procedure in e-healthcare [100]. For example, the IoT will optimize the provision of urban medical services, and this will improve both intra-sectoral and inter-sectoral communications. Patient data are valuable assets that need to be stored and referred to, also in the future. Cloud computing will serve to reduce the human labor required for the storage, retrieval, and interpretation of these records. Cloud computing may also be utilized to construct databases and platforms for medical and health services. Additionally, patients may register remotely through the IoT, and this healthcare access can be provided to 100% of the urban population quite easily [101, 102]. Cloud computing can perform real-time forecasting and evaluation of traffic conditions, analyze traffic development trends, offer trustworthy decision-making resources for traffic management departments, and facilitate road traffic guidance. Based on the current remote monitoring system of traffic, the traffic signal lights are managed and controlled to achieve adequate traffic flow control, and almost eliminate the likelihood of an accident. The standardization and unification of road traffic management may also be achieved by remote traffic control and electronic toll collection, eliminating the problem of slack law enforcement brought on by "human-friendly" management and promoting society's peaceful and steady growth [102]. IoT devices are often widely dispersed across different regions, but they all transmit data to a single central system [109]. As IoT data volumes increase, a corporation may run out of bandwidth and cloud capacity. Additionally, gathering, transmitting, processing, and receiving data at its destination become more time consuming. Furthermore, inefficiencies result from this "delay" especially in businesses where real-time information is crucial for success in the marketplace. With the help of edge computing technologies, the processing can be decentralized and moved closer to the data source [103], by effectively deploying localized computing systems and boosting the processing capability of the IoT devices in the process.

- The sensor-based smart city of the future will be composed of four technological layers [104]. Sensors, which can be automated terminals, wireless and mobile sensors, and network cameras recording images, videos, and sound.

- All these sensors on the city network, connected through the communication infrastructure

- Data collected by the sensors are consolidated, processed, and analyzed to obtain useful information

Time series analysis is possible to understand the trends and patterns, related to energy usage, noise reduction, traffic optimization, and implementation of safety and security measures [105].

4.2 Smart surveillance

Urban lighting systems can double up as surveillance systems in a smart grid [106]. Installing street lighting surveillance technology does not require creating a complete smart grid. In numerous UK cities, a microphone system is installed on streetlights, and the possibility of adding cameras to them is being discussed, at the time of writing. People who live in regions that are vulnerable to crime and terrorism have experienced several changes in their daily routines, which have affected the quality of their lives [107]. Microphones installed on lighting systems to detect angry and suspicious voices, activate cameras connected to the, investigate the sources of the voices further—this is in vogue in New York City for example. As the populations in cities keep increasing, so does the crime rate, as the potential returns for criminal activities are obviously larger in bigger, densely populated cities with more wealthy inhabitants [108, 109]. The standards for installing and employing video surveillance are not usually explicitly and comprehensively defined in legal terms, and this absence of standardization leads to cities in different countries adopting different approaches to avail of this technology to fight crime [110]. Regardless of the weather, intelligent technology can "recognize" faces up to 70 meters [111, 112]. Real-time analysis enables one to react immediately to emergencies.

5. Illustrating with a case

5.1 Smart city model

Expectations of urban inhabitants have changed radically, owing to the realization that ICTs are indeed able to provide them with better services than before. This necessitates better space planning (as urban space is a limited resource) [113]. The effective use of urban space is of paramount importance if transport-related problems are to be solved in a smart city. There is a critical upper limit to the number of vehicles, which can be plying on the roads (again, fixed lengths and thereby fixed traffic capacity) at a given time, without causing congestion. Parking spaces are always at a premium when the fleet of private vehicles in the smart city increases. Thereby, one finds the clever use of underground spaces [114], or the introduction of multi-storeyed parking facilities.

It goes without saying that in developed cities, the creation of a digital matrix of target indicators is crucial as a management tool for a "smart" system and as a description of the current space-planning structure of the city, both above and below the ground. As a result, soft or non-physical assets share a significant capital component with multiple effects in many situations for a smart city. In addition to the advantages offered by the conventional physical infrastructure, they allow a city to implement and mainstream a people-centered strategy [115]. The combination of typical indicators and opportunities adequate to the spatial resources of the city development provides the basis for identifying the development vectors of the system, and the thresholds for their actual implementation—the so-called certainty thresholds. In year-2013, Vienna, Toronto, Paris, New York, London, Tokyo, Berlin, Copenhagen, Hong Kong, Barcelona, Boston, San Francisco, Amsterdam, Karamay, Singapore, Songdo, and Sao Paulo were rated as smart cities [116–118]. Four years later, in 2017, Copenhagen was named as the most technologically advanced one. Cities such as Singapore, Stockholm, Zurich, Boston, Tokyo, San Francisco, Amsterdam, Geneva, and Melbourne were the other 9 in the top-10 list. The next subsection focuses on Singapore in greater detail.

5.2 Singapore as a smart city

Data generated by billions of individuals daily through their usage of modern technologies and social media had made artificial intelligence possible [119]. Singapore has structured its public and private transportation networks, installed smart traffic lights and sensors to measure traffic congestion, introduced smart parking throughout the city, and will soon see the widespread usage of autonomous cars. Despite being a wealthy country, Singapore is well known for its strict social laws. A few years ago, Singapore started an initiative to transition to a smart city. The "whole of nation" approach was adopted to give better assurance of success and experiment with new technologies and concepts of IoT and CPS (Cyber-Physical Systems). Technology maturity, ease of use, and public acceptance were emphasized. All aspects of urban infrastructure, including transportation, telecommunications, healthcare, and resources management, would be encompassed [120]. To date, the whole world is interested in Singapore's emphasis on environment-friendliness in their urban design approach. Technology has enabled a selective application of robotics to either supplement or replaces tedious human labor [121]. The Smart Nation concept aims to use sensors linked to aggregation boxes to gather citywide data digitally. The competent agencies get the data collected on pedestrian activity or traffic volume for analysis and decision making in the delivery of services. The National Research Foundation oversees enhancing Virtual Singapore, a dynamic 3D city model, and a collaborative data platform for planning. Public and commercial businesses can use it to create tools to evaluate ideas and products, such as modeling crowd dispersals from potential sports arenas [121]. The government wants to install solar panels on the roofs of 6000 buildings by 2022, as well as smart and energy-efficient lighting for all public routes [121]. Singapore is now really "smart," thanks to several strategic programs undertaken by the government in close

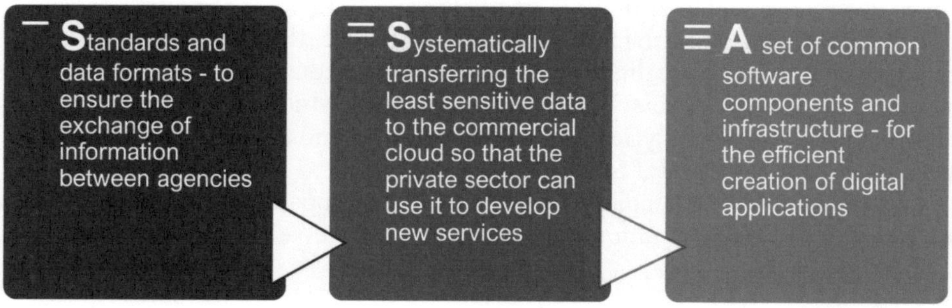

Figure 9.
CODEX platform [122].

collaboration with other stakeholders in the country. New technologies are currently being tested in the city. It relates to "life issues" and the electronic government. They develop a CODEX, a digital platform that swiftly and effectively offers citizens digital services (see **Figure 9**) [122, 123]. Through the Moments of Life initiative, parents can register a new-born, find a kindergarten, or find information about necessary vaccinations, older people can find out about public services, and professionals can find a job.

5.3 Smart public transport and unmanned vehicles

Notwithstanding recent developments, resolving traffic congestion problems has not been easy in Singapore [124]. The government is seeking to address this issue by restricting the number of vehicles on the road and raising costs. However, a successful transportation policy may be achieved with innovative ideas:

- Sensors: With their assistance, traffic can be diverted when sectional congestion is identified.

- Autonomous taxis: The cost of transportation will be lower without a driver.

Over time, people have realized the negative impacts (congestion, noise, and air pollution) of unrestrained use of private vehicles—both four wheelers and two wheelers. There is an increasing predilection for the use of public transportation. Out of 5.6 million people, there were 7.54 million daily bus trips nationwide in 2018, and attempts are constantly made to improve the quality of service provided [125]. To test robotic automobiles, the whole western portion of the nation—more than 1000 kilometers of roads—was made available to businesses. Since last year, locals have used autonomous busses and shuttles, and the nation's first unscrewed taxis debuted in 2016 [126]. Standards for drones have been formalized, to increase the effectiveness of developing and introducing new robots on the road.

5.4 Smart multipurpose lighting

Making a city smart requires investments in high-speed fiber networks, smart city technologies, a strategy for data-sharing and data security, and a master plan for urban growth. The adjective "smart" must perforce also imply "sustainable," "equitable," and "inclusive" [127]. The "Lamp Pole as a Platform (LaaP)" trial initiative was launched by Singapore in 2018, and streetlights were integrated with various sensors, including cameras with facial recognition software. Upscaling to track human activity and monitor air quality, will make it multi-functional and multipurpose. The project is a part of the broader Smart Nation program, which aims to use cutting-edge technology for "crowd analytics," to combat terrorism and augment the safety and security of the inhabitants of Singapore (**Figure 10**).

To date, over 100,000 lampposts in Singapore are equipped with surveillance cameras, which might soon assist law enforcement in identifying people in crowds (**Figure 11**).

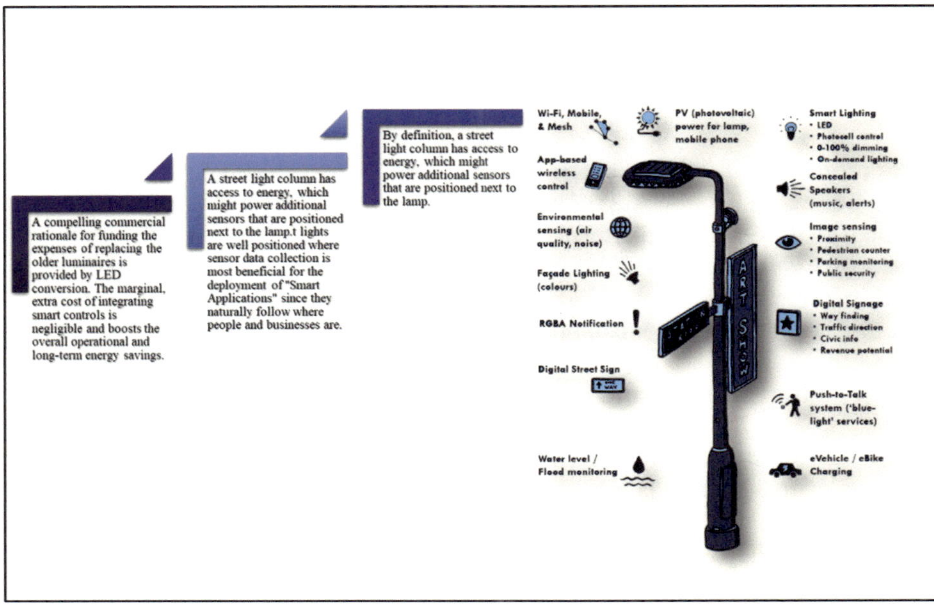

Figure 10.
'Lamp pole as a platform' initiative [128].

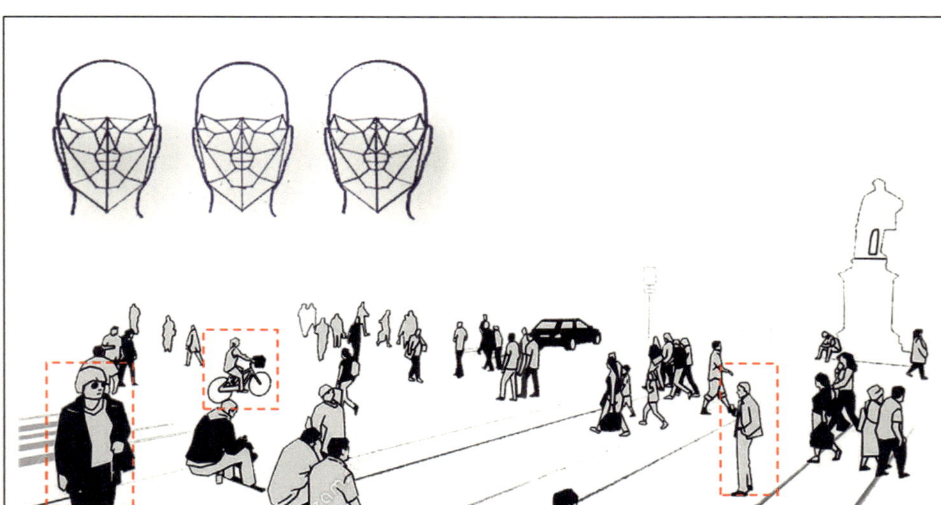

Figure 11.
The technique for identifying faces in crowds.

Such video surveillance can stifle free expression and association, violate the rights to travel and rest, and erode the right to privacy in general [129].

5.5 Education and tech-savviness

In the 1960s, English was mandated as the medium of instruction in schools, and universities started teaching entirely in English. Additionally, the government-sponsored students who intended to study abroad. The bilingualist strategy focusing on speaking English and the native tongues of the three major ethnic groups: Tamil for Indians, Mandarin for the Malay population, and Chinese for the Chinese community, mandates that pupils in schools learn both their respective mother tongues and English [130]. This characterizes a high level of concern and respect for human capital. As part of the Smart Nation Fellowship, specialists from abroad are attracted to Singapore [131]. By analyzing Singapore's recent Smart Nation effort, the authors of this chapter would like to recommend the inclusion of the actual human and embodied work into the 'smart urbanism' paradigm.. Smart Nation Together is an educational course on programming and teaching technologies, including 3D printing and artificial intelligence, designed for children, adults, and the elderly [132].

5.6 Healthy is smart, smart is healthy

A complex housing several research facilities, medical facilities, and entertainment spaces will be constructed in Singapore, by 2030. Health City Novena's master plan aims to create an infrastructure that makes sidewalks, underground parking lots, and parks as comfortable as possible for patients [133]. The developers have made the inanimate infrastructure subservient to the animate entities it is meant to serve—patients, students, guests, and employees of these facilities. Singapore's digital environment makes it possible to both receive and provide services. If someone needs urgent medical assistance, a responder app alerts medical volunteers within a range of a kilometer [134]. Harnessing the benefits of the "enabling aspects of technologies" to the fullest for its citizens, is what has been the hallmark of Singapore's entrenchment as a leading smart city (city-state) in the world [135].

Highly reliable high-speed Internet and the ubiquity of smartphones (some Singaporeans own more than one smartphone) characterize Singapore. This makes it possible to unburden doctors, by organizing remote consultations in cases where the ailment is mild, or the patient query is simply about preventative/prophylactic measures [136].

5.7 Continuous collaborative innovation

Singapore has been a hotbed of innovations, and a crucible for experimentation, for many years now. The government, in its capacity as financier and promoter, works shoulder to shoulder with tech-start-ups and experts in academia and industry, to keep innovating, testing, and implementing briskly.. Singapore's innovation policy while facilitating continuous urban transformation within this city state also has positive spillover effects on other countries in ASEAN, Asia and elsewhere in the world, which can learn from the success of this innovation-powerhouse in South-East Asia [137].

5.8 Digital Singapore enabling resources management

Thousands of cameras and sensors make it possible to create a digital copy of the city. This is necessary to predict various events, from natural disasters to pandemics [138]. In addition, the digital model keeps track of the number of inhabitants, resource consumption, climate change, and many other factors that impact their lives, favorably or otherwise. Resource scarcity (or rather an impending resource scarcity) is directly correlated with the population density of a city. Smart meters and sensors come in handy here, to limit resource consumption in both the residential and commercial sectors. Singapore has incorporated cutting-edge automation technologies—bolstered by ubiquitous sensing and data gathering—to address a host of challenges related to the management of resources [139]. Solar panels are integrated into building facades, to generate electricity for captive use. The NEWater wastewater treatment and recycling system meet 30% of the city's potable-water needs, reducing its dependence on freshwater imports from Malaysia. Modular vertical farms, designed to enable all the plants to have adequate sunlight and water, enable efficient farming, in a country where arable land is at a premium.

6. Results and conclusions

A smart city takes recourse to digital technologies to streamline all "urban operations" to enable quality-of-life improvement for its inhabitants. In addition to promoting social well-being, it also has a beneficial impact on industrial development and economic growth. Businesses are impacted favorably by digitization, as it gives them an attractive return on investment. Emergencies necessitating quick remedial measures can be effectively managed if smart technologies are availed of. The IT infrastructure of a smart city is comprised of several networked computers, controllers, sensors, and devices. These subsystems gather huge volumes of data that need to be stored, crunched, and transmitted. The city administration can communicate directly with the inhabitants (as well as with all the urban infrastructure elements) through the IoT. This will enable the administrators to monitor and measure in real-time and combine proactive and reactive approaches to adapt and evolve and keep enhancing the quality of life of the city's inhabitants. Building a smart city calls for comprehensive perception, ubiquitous interconnection, ubiquitous computing, and integrated applications through new-generation IT applications like the IoT and cloud computing, represented by mobile technology.

Wikis, social networks, and complete integration techniques must be used in smart cities from the standpoint of social development to actualize a knowledge society characterized by using creativity and innovation which is free, collaborative, and benefits the masses. For instance, smart (adjustable) street lighting can conserve a lot of energy. Smart traffic lights can autonomously select the mode of operation depending on their analysis of the flow of vehicles and the traffic scenario. All public locations with video cameras and emergency call panels serve to ensure security. Real-time information—about various aspects of the lives of the citizens, like healthcare, utilities, security, and transportation, inter alia—streamed from the cameras is gathered and examined. Smart houses and buildings are also a component of developing smart cities.

Managing urban infrastructure, such as transportation, education, healthcare, housing and community services, security, requires the integration of several ICTs.

The overarching goal of developing a "smart city" is to augment the standard of living of its citizens by employing urban informatics technologies to enhance the effectiveness of the services provided. For a qualitative improvement in the level of security, a move to proactive actions that enable crime prediction and resource allocation planning is required. By examining historical data on antecedents, it is possible to develop risk profiles.

A smart city is an advanced kind of "urban informatization" that completely utilizes the new generation of ICTs in all areas of urban life. Ideas for regional development like e-government, intelligent transportation, and smart grids commonly merge with smart cities. Smart cities must necessarily have other attributes such as "eco-friendly" and "low-carbon."

While technological innovations aided by ICTs are looked upon as solutions for all challenges by some techno-optimists, it must be borne in mind that "smartness" is incomplete without human-networks-building and human capital development. Smart cities, after all, must be "of the people," "by the people," and "for the people," with technology being just a humble servant of the masses.

Author details

Amjad Almusaed[1*] and Asaad Almssad[2]

1 Department of Construction Engineering and Lighting Science, Jonkoping University, Sweden

2 Faculty of Health, Science and Technology, Karlstad University, Sweden

*Address all correspondence to: amjad.al-musaed@ju.se

References

[1] Karen C, Burak G, Lucy R. Global forecasts of urban expansion to 2030 and direct impacts on biodiversity and carbon pools. PNAS. 2012;**109**(40):16083-16088. DOI: 10.1073/pnas.1211658109

[2] Yitmen I, Almusaed A, et al. ANP model for evaluating the performance of adaptive façade systems in complex commercial buildings. Engineering, Construction and Architectural Management Journal. 2021;**29**(1):431-455. DOI: 10.1108/ECAM-07-2020-0559

[3] Sancino A, Hudson L. Leadership in, of, and for smart cities – Case studies from Europe. America, and Australia, Public Management Review. 2020;**22**(5):701-725. DOI: 10.1080/14719037.2020.1718189

[4] Almssad AA. Environmental reply to vernacular habitat conformation from vast areas of Scandinavia. Renewable & Sustainable Energy Reviews Journal. 2015;**48**:825-834. DOI: 10.1016/j.rser.2015.04.013

[5] McNeill D. Global firms and smart technologies: IBM and the reduction of cities. Transactions. 2015;**40**(4):562-574. DOI: 10.1111/tran.12098

[6] Gul S, Yang H, Ahmad AW, Lee C. Energy-efficient intelligent street lighting system using traffic-adaptive control. IEEE Sensors Journal. 2016;**16**(13):1-1. DOI: 10.1109/JSEN.2016.2557345

[7] Almusaed A, Almssad A. Lessons from the world sustainable housing (past experiences, current trends, and future strategies). In: Sustainable Housing. London, UK, London: IntechOpen; 2021. DOI: 10.5772/intechopen.100533 Available from: https://www.intechopen.com/online-first/79055

[8] Almusaed A, Almssad A. Introductory chapter: Sustainable housing – Introduction to the thematic area. In: Sustainable Housing. London: IntechOpen; 2021. DOI: 10.5772/intechopen.101968 Available from: https://www.intechopen.com/chapters/79962

[9] Pardini K, Rodrigues JJPC, Diallo O, Das AK, de Albuquerque VHC, Kozlov SA. A smart waste management solution geared towards citizens. Sensors. 2020;**20**(8):2380. DOI: 10.3390/s20082380

[10] Ammara U, Rasheed K, Mansoor A, Al-Fuqaha A, Qadir J. Smart cities from the perspective of systems. Systems. 2022;**10**(3):77. DOI: 10.3390/systems10030077

[11] Khalid A, Ali A, Amiya B. The role of smart government characteristics for enhancing UAE's public service quality. International Journal on Emerging Technologies. 2019;**10**(1a):01-07

[12] Alexander V, Irina V. Smart cities in the far north in 2038. Ten factors that will influence the development of smart cities the coming twenty years, conference. In: 5th Sgem International Multidisciplinary Scientific Conferences on Social Sciences and Arts sgem2018. Sofia, Bulgaria: SGEM Social Sciences and Arts & Humanities; 2018. DOI: 10.5593/sgemsocial2018/5.2/S19.026

[13] Belli L, Cilfone A, Davoli L, Ferrari G, Adorni P, Di Nocera F, et al. IoT-enabled smart sustainable cities: Challenges and approaches. Smart Cities.

2020;**3**(3):1039-1071. DOI: 10.3390/smartcities3030052

[14] Almusaed A, Almssad A. Introductory chapter: Overview of sustainable cities, theory and practices. In: Almusaed A, Almssad A, editors. Sustainable Cities. United Kingdom: InTech; 2019. DOI: 10.5772/intechopen.82632 Available from: https://www.intechopen.com/chapters/64752

[15] Caragliu A, Del Bo C, Nijkamp P. Smart Cities in Europe, Serie Research Memoranda 0048. VU University Amsterdam: Faculty of Economics, Business Administration and Econometrics; 2009

[16] Joss S, Cook M, Dayot Y. Smart cities: Towards a new citizenship regime? A discourse analysis of the British Smart City standard. Journal of Urban Technology. 2017;**24**(4):29-49. DOI: 10.1080/10630732.2017.1336027

[17] Frost & Sullivan: Global Smart Cities market to reach US$1.56 trillion by 2020. Available from: http://ww2.frost.com/news/press-releases/frost-sullivan-global-smart-cities-market-reach-us156-trillion-2020. [Accessed: May 18, 2022]

[18] Nižetić S, Šolić P, López-de-Ipiña González-de-Artaza D, Patrono L. Internet of things (IoT): Opportunities, issues and challenges towards a smart and sustainable future. Journal of Cleaner Production. 2020;**274**:122877. DOI: 10.1016/j.jclepro.2020.122877

[19] Hannah Ritchie, Max Roser."Urbanization". Published online at OurWorldInData.org. 2018. Available from: https://ourworldindata.org/urbanization

[20] Satterthwaite D. Will most people live in cities? BMJ. 2000;**321**:1143. DOI: 10.1136/bmj.321.7269.1143

[21] Bauer M, Sanchez L, Song J. IoT-enabled smart cities: Evolution and outlook. Sensors. 2021;**21**(13):4511. DOI: 10.3390/s21134511

[22] Morawska L et al. Applications of low-cost sensing technologies for air quality monitoring and exposure assessment: How far have they gone? Environmental International. 2018;**116**:286-299. DOI: 10.1016/j.envint.2018.04.018

[23] Almusaed A, Almssad A, Alasadi A. Analytical interpretation of energy efficiency concepts in the housing design. Journal of Building Engineering. 2019;**21**:254-266. DOI: 10.1016/j.jobe.2018.10.026

[24] Buzura S, Iancu B, Dadarlat V, Peculea A, Cebuc E. Optimizations for energy efficiency in software-defined wireless sensor networks. Sensors. 2020;**20**(17):4779. DOI: 10.3390/s20174779

[25] Pandya P et al. Dream City–a cutting edge worldwide city. International Journal for Research in Applied Science & Engineering Technology (IJRASET). 2018;**6**(III). DOI: 10.22214/ijraset.2018.3404

[26] Satterthwaite D, McGranahan G, Tacoli C. Urbanization and its implications for food and farming. Philosophical Transactions of the Royal Society of London. Series B, Biological Sciences. 2010;**365**(1554):2809-2820. DOI: 10.1098/rstb.2010.0136. PMID: 20713386; PMCID: PMC2935117

[27] Walton JH. "Chapter 3 the Ancient Cosmological Cognitive Environment".

Genesis 1 as Ancient Cosmology. University Park, USA: Penn State University Press; 2021. pp. 23-121. DOI: 10.1515/9781575066547-005

[28] Almusaed A, Almssad A. Building materials in eco-energy houses from Iraq and Iran. Case Studies in Construction Materials Journal. 2015;**2**:42-54. DOI: 10.1016/j.cscm.2015.02.001

[29] Kotchoubey B. Human consciousness: Where is it from and what is it for. Frontiers in Psychology. 2018;**9**:567. DOI: 10.3389/fpsyg.2018.00567

[30] Jacinta F, Billie G, Lisa W, Matthew K. Creating a sense of community: The role of public space. Journal of Environmental Psychology. 2012;**32**(4):401-409. DOI: 10.1016/j.jenvp.2012.07.002

[31] Almusaed A, Almssad A, Karim N. Appreciative inquiry approach upon biophilic factors within school spaces design from Scandinavia. Advances in Civil Engineering. 2022;**2022**:8545787. DOI: 10.1155/2022/8545787

[32] Almusaed A, Almssad A. Sustainable Cities - Authenticity, Ambition and Dream. London, UK, The United Kingdom: IntechOpen; 2019. DOI: 10.5772/73410

[33] Simon EB, John K, Mattias K. Compact city planning and development: Emerging practices and strategies for achieving the goals of sustainability. Developments in the Built Environment. 2020;**4**:100021. DOI: 10.1016/j.dibe.2020.100021

[34] Almusaed A, Alasadi A, Almssad A. A research on the biophilic concept upon school's design from hot climate: A case study from Iraq. Advances in Materials Science and Engineering. 2022;**2022**:12. Article 7994999. DOI: 10.1155/2022/7994999

[35] Mouratidis K. Urban planning, and quality of life: A review of pathways linking the built environment to subjective well-being. Cities. 2021;**115**:103229. DOI: 10.1016/j.cities.2021.103229

[36] Prieto Curiel R, Bishop SR. Fear of crime: The impact of different distributions of victimisation. Palgrave Communication. 2018;**4**:46. DOI: 10.1057/s41599-018-0094-8

[37] Mills DE, Izadgoshasb I, Pudney SG. Smart City collaboration: A review and an agenda for establishing sustainable collaboration. Sustainability. 2021;**13**(16):9189. DOI: 10.3390/su13169189

[38] Sharifi A, Allam Z, Feizizadeh B, Ghamari H. Three decades of research on smart cities: Mapping knowledge structure and trends. Sustainability. 2021;**13**(13):7140. DOI: 10.3390/su13137140

[39] David T, Kalle L, Carsten S. Research commentary—Digital infrastructures: The missing IS research agenda. Information Systems Research;**21**(4):748-759

[40] Serrano W. Digital Systems in Smart City and Infrastructure: Digital as a service. Smart Cities. 2018;**1**(1):134-154. DOI: 10.3390/smartcities1010008

[41] Juan CM, Michel A, Jonas H. Development Dynamics of Digital Infrastructure and Organization: The Case of Global Payments Innovation. Munich: Development Dynamics of Digital Infrastructure and Organization, Fortieth International Conference on Information Systems; 2019

[42] Ted S, Ulrika HW, Tomas B. Digital transformation: Five recommendations for the digitally conscious firm. Business

Horizons. 2020;**63**(6):825-839. DOI: 10.1016/j.bushor.2020.07.005

[43] Wayne F, Ramiro M. How technology is changing work and organizations. Annual Review of Organizational Psychology and Organizational Behaviour. 2016;**3**:349-375. DOI: 10.1146/annurev-orgpsych-041015-062352

[44] Javed A et al. Future Smart Cities: Requirements, Emerging Technologies, Applications, Challenges, and Future Aspects. Preprint: TechRxiv; a 2021. 10.36227/techrxiv.14722854.v1

[45] Ismagilova E, Hughes L, Rana NP, et al. Security, privacy and risks within smart cities: Literature review and development of a Smart City interaction framework. Information Systems Frontiers. 2020;**24**:393-414. DOI: 10.1007/s10796-020-10044-1

[46] Almusaed A, Almssad A. Housing policy matters. In: Almusaed A, Almssad A, editors. Housing. United Kingdom: InTech; 2018. DOI: 10.5772/intechopen.81622 Available from https://www.intechopen.com/chapters/64126

[47] Camargo F et al. Towards a new model of smart cities in emerging countries. Academy of Strategic Management Journal. 2021;**20**(6):1-20

[48] Bellini P, Nesi P, Pantaleo G. IoT-enabled smart cities: A review of concepts, frameworks and key technologies. Applied Sciences. 2022;**12**(3):1607. DOI: 10.3390/app12031607

[49] Chew Michael YL, Teo Evelyn AL, Shah Kwok W, Kumar Vishal, Hussein Ghassan F. valuating the roadmap of 5G technology implementation for smart building and facilities Management in Singapore. Sustainability. 2020;**12**(24): 10259. DOI:10.3390/su122410259

[50] Toma's S, Daniel C. Ranasinghe Mark H, Duncan MF. Adding sense to the internet of things; an architecture framework for smart object systems. Personal and Ubiquitous Computing. 2012;**16**:291-308. DOI: 10.1007/s00779-011-0399-8

[51] Lynn T, Endo PT, Ribeiro AMNC, Barbosa GBN, Rosati P. The internet of things: Definitions, key concepts, and reference architectures. In: Lynn T, Mooney J, Lee B, Endo P, editors. The Cloud-to-Thing Continuum. Palgrave Studies in Digital Business & Enabling Technologies. Cham: Palgrave Macmillan; 2020. DOI: 10.1007/978-3-030-41110-7_1

[52] Mehami M, Nawi RY. Smart automated guided vehicles for manufacturing in the context of industry 4.0. Procedia Manufacturing. 2018;**26**:1077-1086. DOI: 10.1016/j.promfg.2018.07.144

[53] Oughton EJ, Lehr W, Katsaros K, Selinis I, Bubley D, Kusuma J. Revisiting wireless internet connectivity: 5G vs Wi-fi 6. Telecommunications Policy. 2021;**45**(5):102127. DOI: 10.1016/j.telpol.2021.102127

[54] Mahamuni A. Internet of things, machine learning, and artificial intelligence in the modern supply chain and transportation. Defense Transportation Journal. 2018;**74**(1):14-17 Available from: https://www.jstor.org/stable/26430583

[55] Darling-Hammond L, Flook L, Cook-Harvey C, Barron B, Osher D. Implications for educational practice of the science of learning and development. Applied Developmental Science.

2020;**24**(2):97-140. DOI: 10.1080/10888691.2018.1537791

[56] Mensah Justice, Casadevall Sandra Ricart. Sustainable development: Meaning, history, principles, pillars, and implications for human action: Literature review. Cogent Social Sciences. 2019;**5**:1. DOI: 10.1080/23311886.2019.1653531

[57] Davenport T, Guha A, Grewal D, et al. How artificial intelligence will change the future of marketing. Journal of the Academy of Marketing Science. 2020;**48**:24-42. DOI: 10.1007/s11747-019-00696-0

[58] Yadav A et al. Adaptive traffic management system using IoT and machine learning. International Journal of Scientific Research in Science, Engineering and Technology. 2019;**6**(1):216-229

[59] Baharudin N, Mansur TN, Ali R, Sobri N. Smart lighting system control strategies for commercial buildings: A review. International Journal of Advanced Technology and Engineering Exploration. 2021;**8**(74):45-53. DOI: 10.19101/IJATEE. 2020.S2762173

[60] Almusaed A. Biophilic and Bioclimatic Architecture, Analytical Therapy for the Next Generation of Passive Sustainable Architecture. London Limited, London, UK: Springer-Verlag; 2011. pp. 202-230. Available from: https://www.springer.com/gp/book/9781849965330

[61] Hancke GP et al. The role of advanced sensing in smart cities. Sensors (Basel, Switzerland). 2012;**13**(1):393-425. DOI: 10.3390/s130100393

[62] Enholm IM, Papagiannidis E, Mikalef P, et al. Artificial intelligence and business value: A literature review. Information Systems Frontiers. 2021:1-26. DOI: 10.1007/s10796-021-10186-w

[63] Caprotti F, Liu D. Platform urbanism and the Chinese smart city: The co-production and territorialisation of Hangzhou City brain. GeoJournal. 2022;**87**:1559-1573. DOI: 10.1007/s10708-020-10320-2

[64] Nellore K, Hancke GP. Traffic management for emergency vehicle priority based on visual sensing. Sensors (Basel). 2016;**16**(11):1892. DOI: 10.3390/s16111892. PMID: 27834924; PMCID: PMC5134551

[65] Chen Y et al. Using 5G in smart cities: A systematic mapping study. Intelligent Systems with Applications. 2022;**14**:200065. DOI: 10.1016/j.iswa.2022.200065

[66] Jofre S. Strategic Management: The Theory and Practice of Strategy in (Business) Organizations. Kgs. Lyngby: DTU Management; 2011. p. 1

[67] Nimra K, Marina E. The use of biometric technology at airports: The case of customs and border protection (CBP). International Journal of Information Management Data Insights. 2021;**1**(2):100049. DOI: 10.1016/j.jjimei.2021.100049

[68] Heratha HMKKMB, Mamta M. Adoption of artificial intelligence in smart cities: A comprehensive review. International Journal of Information Management Data Insights. 2022;**2**(1):100076. DOI: 10.1016/j.jjimei.2022.100076

[69] Chander B, Pal S, De D, Buyya R. Artificial intelligence-based internet of things for industry 5.0. In: Pal S, De D, Buyya R, editors.

Artificial Intelligence-Based Internet of Things Systems. Internet of Things. Cham: Springer; 2022. DOI: 10.1007/978-3-030-87059-1_1

[70] Myeong S, Kim Y, Ahn MJ. Smart City strategies—Technology push or culture pull? A case study exploration of Gimpo and Namyangju, South Korea. Smart Cities. 2021;**4**(1):41-53. DOI: 10.3390/smartcities4010003

[71] Alzahrani NM, Alfouzan FA. Augmented reality (AR) and cyber-security for smart cities—A systematic literature review. Sensors. 2022;**22**(7):2792. DOI: 10.3390/s22072792

[72] Suchita J, Sujata J. Role of augmented reality applications for smart city planning. International Journal of Innovative Technology and Exploring Engineering (IJITEE). 2019;**8**(9S2):2278-3075

[73] Yewande M. Virtual reality as a tool for learning: The past, present and the prospect. Journal of Applied Learning & Teaching. 2020;**3**(2):51-58. DOI: 10.37074/jalt.2020.3.2.10

[74] Sanche M. Virtual interactive innovations applied for digital urban transformations. Mixed approach, Future Generation Computer Systems. 2019;**91**:371-381. DOI: 10.1016/j.future.2018.08.016

[75] Ompad DC, Galea S, Vlahov D. Urbanicity, urbanization, and the urban environment. In: Macrosocial Determinants of Population Health. New York, NY: Springer; 2007. DOI: 10.1007/978-0-387-70812-6_3

[76] Lv Z, Yin T, Zhang X, Song H, Chen G. Virtual reality Smart City based on WebVRGIS. IEEE Internet of Things Journal. 2016;**3**(6):1015-1024. DOI: 10.1109/JIOT.2016.2546307

[77] Kong L, Woods O. The ideological alignment of smart urbanism in Singapore: Critical reflections on a political paradox. Urban Studies. 2018;**55**(4):679-701. DOI: 10.1177/0042098017746528

[78] Lee A, Lee K-W, Kim K-H, Shin S-W. A geospatial platform to manage large-scale individual mobility for an urban digital twin platform. Remote Sensing. 2022;**14**(3):723. DOI: 10.3390/rs14030723], 10.3390/rs14030723]

[79] Panagiotis S, Dimitris M. Design and operation of production networks for mass personalization in the era of cloud technology. In: Digital Twins in Industry 4.0. Swesland: Elsevier B.V.; 2022. pp. 277-316. DOI: 10.1016/B978-0-12-823657-4.00010-5

[80] Mourtzis D. Simulation in the design and operation of manufacturing systems: State of the art and new trends. International Journal of Production Research. 2020;**58**(7):1927-1949. DOI: 10.1080/00207543.2019.1636321

[81] Microsoft blog. Available from: https://blogs.microsoft.com/iot/2018/09/24/announcing-azure-digital-twins-create-digital-replicas-of-spaces-and-infrastructure-using-cloud-ai-and-iot/

[82] Botín-Sanabria DM, Mihaita A-S, Peimbert-GarcíaRE,Ramírez-MorenoMA, Ramírez-Mendoza RA, Lozoya-Santos JJ. Digital twin technology challenges and applications: A comprehensive review. Remote Sensing. 2022;**14**(6):1335. DOI: 10.3390/rs14061335

[83] Singh M, Fuenmayor E, Hinchy EP, Qiao Y, Murray N, Devine D. Digital

twin: Origin to future. Applied System Innovation. 2021;**4**(2):36. DOI: 10.3390/asi4020036

[84] Boulouf A, Sedqui A, Chater Y. Connecting maintenance management and industry 4.0 technology. Academy of. Strategic Management Journal. 2022;**21**(3):1-20

[85] Fan X, Weisheng L, Zhe C, Christopher JW. From LiDAR point cloud towards digital twin city: Clustering city objects based on gestalt principles. ISPRS Journal of Photogrammetry and Remote Sensing. 2020;**167**:418-431. DOI: 10.1016/j.isprsjprs.2020.07.020

[86] Gohari A, Ahmad AB, Rahim RBA, Supa'at ASM, Razak SA, Gismalla MSM. Involvement of surveillance drones in smart cities: A systematic review. IEEE Access. 2022;**10**:56611-56628. DOI: 10.1109/ACCESS.2022.3177904

[87] Jacek K, Beata Ś. Improvements in the quality of courier delivery. International Journal for Quality Research;**10**(2):355-372. DOI: 10.18421/IJQR10.02-08

[88] Muhammad AK, Bilal AA, Engr AS, Inam UK. Drones for good in smart cities: A review. In: Conference: International Conference on Electrical, Electronics, Computers, Communication, Mechanical and Computing (EECCMC) 28th & 29 th January 2018. India: IEEE; 2018

[89] Restás Á. Drone applications fighting COVID-19 pandemic—Towards good practices. Drones. 2022;**6**(1):15. DOI: 10.3390/drones6010015

[90] Aleksandar M, Aca R, Marko R. Use of drons In operations in the urban environment. In: V International Scientific Conference Safety and Crisis Management – Theory and Practice Safety for the Future – SecMan. Belegrad: Regional Association for Security and Crisis Management; 2019

[91] Klimkowska A, Lee I, Choi K. POSSIBILITIES OF UAS FOR MARITIME MONITORING. In: The International Archives of the Photogrammetry, Remote Sensing and Spatial Information Sciences. Vol. XLI-B1. Prague, Czech Republic: XXIII ISPRS Congress, 12-19 July 2016; 2016. DOI: 10.5194/isprsarchives-XLI-B1-885-2016

[92] Burke C, Wich S, Kusin K, et al. Thermal-drones as a safe and reliable method for detecting subterranean peat fires. Drones. 2019;**3**(1):23. DOI: 10.3390/drones3010023

[93] Cai H, Xu B, Jiang L, Vasilakos AV. IoT-based big data storage systems in Cloud Computing: Perspectives and challenges. IEEE Internet of Things Journal. 2017;**4**(1):75-87. DOI: 10.1109/JIOT.2016.2619369

[94] Ray PP. A survey on internet of things architectures. Journal of King Saud University - Computer and Information Sciences. 2018;**30**(3):291-319. DOI: 10.1016/j.jksuci.2016.10.003

[95] Alam T. Cloud-based IoT applications and their roles in smart cities. Smart Cities. 2021;**4**(3):1196-1219. DOI: 10.3390/smartcities4030064

[96] Almusaed A. A general reading process on landscape architecture. In: Almusaed A, Almssad A, editors. Landscape Architecture, Chapter 1. United Kingdom: InTech; 2018. DOI: 10.5772/intechopen.77971 Available from: https://www.intechopen.com/

books/landscape-architecture-the-sense-of-places-models-and-applications/introductory-chapter-a-general-reading-process-on-landscape-architecture

[97] Black D, Black J. A review of the urban development and transport impacts on public health with particular reference to Australia: Trans-disciplinary research teams and some research gaps. International Journal of Environmental Research and Public Health. 2009;**6**(5):1557-1596. DOI: 10.3390/ijerph6051557

[98] Almusaed A. Introductory chapter: A general reading process on landscape architecture. In: Almusaed A, Almssad A, editors. Sustainable Buildings, Chapter 1. United Kingdom: InTech; 2018. DOI: 10.5772/intechopen.77971 Available from: https://www.intechopen.com/chapters/61575

[99] Bibri SE. On the sustainability of smart and smarter cities in the era of big data: An interdisciplinary and transdisciplinary literature review. Journal of Big Data. 2019;**6**:25. DOI: 10.1186/s40537-019-0182-7

[100] Devadass L, Sekaran SS, Thinakaran R. Cloud computing in healthcare. International Journal of Students' Research in Technology & Management. 2017;**5**(1):25-31. DOI: 10.18510/ijsrtm.2017.516

[101] Dow-Fleisner SJ, Seaton CL, Li E, et al. Internet access is a necessity: A latent class analysis of COVID-19 related challenges and the role of technology use among rural community residents. BMC Public Health. 2022;**22**:845. DOI: 10.1186/s12889-022-13254-1

[102] Alsaawy Y, Alkhodre A, Abi Sen A, Alshanqiti A, Bhat WA, Bahbouh NM. A comprehensive and effective framework for traffic congestion problem based on the integration of IoT and data analytics. Applied Sciences. 2022;**12**(4):2043. DOI: 10.3390/app12042043

[103] Atieh AT. The next generation cloud technologies: A review on distributed cloud, fog and edge computing and their opportunities and challenges. ResearchBerg Review of Science and Technology. 2021;**1**(1):1-15. Available from: https://researchberg.com

[104] Okafor CC et al. IOP Conf. Ser. Materials Science and Engineering. 2021;**1107**:012228

[105] Chen L, Han P. The construction of a Smart City energy efficiency management system oriented to the mobile data aggregation of the internet of things. Complexity. 2021;**2021**:9988282. DOI: 10.1155/2021/9988282

[106] Cellucci L et al. Urban lighting project for a small town: Comparing citizens and authority benefits. Sustainability. 2015;**7**(10):14230-14244. DOI: 10.3390/su71014230

[107] Algahtany M, Kumar L. A method for exploring the link between urban area expansion over time and the opportunity for crime in Saudi Arabia. Remote Sensing. 2016;**8**(10):863. DOI: 10.3390/rs8100863

[108] Malik AA. Urbanization and crime: A relational analysis. IOSR Journal Of Humanities And Social Science (IOSR-JHSS). 2016;**21**(1):68-74. DOI: 10.9790/0837-21146874

[109] Martínez R, Rosenfeld R, Mares D. Social disorganization, drug market activity, and Neighborhood violent crime. Urban Affairs Reviews. 2008;**43**(6):846-874. DOI:

10.1177/1078087408314774. PMID: 19655037; PMCID: PMC2719901

[110] Socha R, Kogut B. Urban video surveillance as a tool to improve security in public spaces. Sustainability. 2020;**12**(15):6210. DOI: 10.3390/su12156210

[111] Dharavath K, Talukdar FA, Laskar RH, Dey N. Face recognition under dry and wet face conditions. In: Dey N, Santhi V, editors. Intelligent Techniques in Signal Processing for Multimedia Security. Studies in Computational Intelligence. Vol. 660. Cham: Springer; 2017. DOI: 10.1007/978-3-319-44790-2_12

[112] Alahakoon D, Nawaratne R, Xu Y, et al. Self-building artificial intelligence and machine learning to empower big data analytics in smart cities. Information Systems Frontiers. 2020:1-20. DOI: 10.1007/s10796-020-10056-x

[113] Peter C et al. Digital transformation: A multidisciplinary reflection and research agenda. Journal of Business Research. 2021;**122**:889-901. DOI: 10.1016/j.jbusres.2019.09.022

[114] Shoup D, editor. Parking and the City. 1st ed. New York: Routledge; 2018. DOI: 10.4324/9781351019668

[115] Wataya E, Shaw R. Soft assets consideration in smart and resilient city development. Smart Cities. 2022;**5**(1):108-130. DOI: 10.3390/smartcities5010007

[116] Sheina S et al. "Smart City": Comfortable living environment. IOP Conference Series: Materials Science and Engineering. **463**(3)

[117] Begić H, Galić M. A systematic review of construction 4.0 in the context of the BIM 4.0 premise. Buildings. 2021;**11**(8):337. DOI:10.3390/buildings11080337

[118] Almusaed A et al. Coherent Investigation on a Smart Kinetic Wooden Façade Based on Material Passport Concepts and Environmental Profile Inquiry. Vol. 2021. Switzerland: MDPI Energy in Construction and Building Materials; 2021. DOI: 10.3390/ma14143771

[119] Mahrez Z, Sabir E, Badidi E, Saad W, Sadik M. Smart urban mobility: When mobility systems meet smart data. IEEE Transactions on Intelligent Transportation Systems. 2022;**23**(7):6222-6239. DOI: 10.1109/TITS.2021.3084907

[120] Chia ES. Singapore's smart nation program — Enablers and challenges. 2016 11th System of Systems Engineering Conference (SoSE). 2016:1-5. DOI: 10.1109/SYSOSE.2016.7542892

[121] Manase P. Singapore model of smart city: A solution to growing urbanization. International Journal of Research in Social Sciences. 2018;**8**(1):768-782

[122] Sønderlund AL, Smith JR, Hutton C, Kapelan Z. Using smart meters for household water consumption feedback: Knowns and unknowns. Procedia Engineering. 2014;**89**:990-997. DOI: 10.1016/j.proeng.2014.11.216

[123] Loconto AM, Poisot AS, Santacoloma P. Innovative markets for sustainable agriculture: How innovations in market institutions encourage sustainable agriculture in developing countries. Rome, Italy: Food and Agriculture Organization of the United Nations and Institut National de la Recherche Agronomique. 2016:390. ISBN: 978-92-5-109327-6

[124] Neuman M. The compact city fallacy. Journal of Planning Education and Research. 2005;**25**(1):11-26. DOI: 10.1177/0739456X04270466

[125] Huu DN, Ngoc VN. Analysis study of current transportation status in Vietnam's urban traffic and the transition to electric two-wheelers mobility. Sustainability. 2021;**13**(10):5577. DOI: 10.3390/su13105577

[126] Taeihagh A, Lim HSM. Governing autonomous vehicles: Emerging responses for safety, liability, privacy, cybersecurity, and industry risks. Transport Reviews. 2019;**39**(1):103-128. DOI: 10.1080/01441647.2018.1494640

[127] Johnston K. A comparison of two smart cities: Singapore & Atlanta. Journal of Comparative, Urban Law and Policy. 2019;**3**(1):191-207

[128] Available from: https://cities-today.com/a-quarter-of-streetlights-could-be-smart-by-2030/

[129] Slobogin, Christopher. Public Privacy: Camera Surveillance of Public Places and The Right to Anonymity. Available from: https://ssrn.com/abstract=364600

[130] Tsung LTH, Cruickshank K. Mother tongue and bilingual minority education in China. International Journal of Bilingual Education and Bilingualism. 2009;**12**(5):549-563. DOI: 10.1080/13670050802209871

[131] Docquier F. The brain drains from developing countries. IZA World of Labor. 2014:31. DOI: 10.15185/izawol.31

[132] Willems T, Graham C. The imagination of Singapore's smart nation as digital infrastructure: Rendering (digital) work invisible. Asian science, technology and society: An International Journal. 2019;**13**(4):511-536. DOI: 10.1215/18752160-8005194

[133] Tixier N, Fiori S, Assefa I, Cifuentes C, Mcoisans J, et al. Bogotá: Case Study: Research 2009-2010. [Research Report] 76. Cresson: Rafael Vinoly Architects; 2010. p. 251

[134] Kjærup M, Elsborg M, Skov MB, Bruun AR. Available Anytime Anywhere: Investigating Mobile Volunteer Responders for Out of Hospital Cardiac Arrest. CHI '21 Proceedings of the 2021 CHI Conference on Human Factors in Computing Systems. 2021:1-13. Article 647. DOI: 10.1145/3411764.3445208

[135] Sharifi A, Khavarian-Garmsir AR. The COVID-19 pandemic: Impacts on cities and major lessons for urban planning, design, and management. Science of the Total Environment. 2020;**749**:142391. DOI: 10.1016/j.scitotenv.2020.142391

[136] Kelly DV, Young S, Phillips L, Clark D. Patient attitudes regarding the role of the pharmacist and interest in expanded pharmacist services. Canadian Pharmacists Journal (Ott). 2014;**147**(4):239-247. DOI: 10.1177/1715163514535731. PMID: 25360150; PMCID: PMC4212442

[137] Anvuur A, Kumaraswamy M. Making PPPs work in developing countries: Overcoming common challenges. In: CIB W107 Construction in Developing Countries International Symposium "Construction in Developing Economies: New Issues and Challenges" January 18th – 20th; 2006 – Santiago, Chile. Netherland: CIB; 2006

[138] Harrison CG, Williams PR. A systems approach to natural disaster resilience. Simulation Modelling

Practice and Theory. 2016;**65**:11-31. DOI: 10.1016/j.simpat.2016.02.008

[139] Hancke GP, Silva Bde C, Hancke GP Jr. The role of advanced sensing in smart cities. Sensors (Basel). 2012;**13**(1):393-425. DOI: 10.3390/s130100393. PMID: 23271603; PMCID: PMC3574682

Chapter 2

Embracing Human Complexity in Service Design for Inclusive and Sustainable Smart Cities

Margherita Pillan

Abstract

According to the UN Agenda 2030, the sustainable development of cities is aimed at innovation for social, environmental, and economical progress. The goal is the development of services and socio-technical systems apt to conjugate inclusivity) with economical and ambient sustainability. To produce social progress, the innovation of infrastructures and services should match the diversified needs of the contemporary multicultural environments and be designed to favor the change of behavior of citizens toward more convenient and sustainable lifestyles. The chapter discusses the importance of the active contribution of citizens to achieving the objectives of the UN Agenda 2030. It argues the opportunity to include scientific theories on human complexity in university training for sustainable service design and proposes the theories of Design for Behavior Change as a valuable conceptual tool. Finally, the chapter focuses on the general value of considering gender perspectives in the design of smart services and systems to optimize satisfaction and adherence to services.

Keywords: sustainability, United Nations agenda 2030, service design, design for behavior change, gender issues

1. Introduction

Smart cities are paradigms for the development of urban environments and the contemporary expression of the eternal human strive toward the perfect settlement.

In human history, the ambition of conceiving ideal urban organizations is very old, and the dreams of the perfect city always reflected the culture, values, and beliefs of the philosopher and architect generating them. The vision of future urban settlements is animated by the intention of progress, by the aspiration of achieving an ideal condition, and overcoming problems and critical situations. Each proposition for urban development reflects, in explicit or implicit ways, a system of values and priorities and a vision of the relationship between humans and the natural environment. From Plato to Leon Battista Alberti in the Italian Renaissance to Le Corbusier, the study of the paradigms proposed for the cities of the future produces a representation of the human aspiration for improvement according to the values of their time and of the struggles for the realization of visions through attempts, errors, and improvements [1].

Plato's idea of roles in society was influenced by the assumption of the natural social distinction between freemen and slaves and by the importance paid to military protection; the wish to contribute to restoring people's dignity inspired Gropius's work [2]; contemporary theories on social innovation express the convincement that people and communities are main stakeholders in the invention of solutions for sustainable and desirable development [3].

The primary purpose of smart cities is the creation of environments and socio-technical ecosystems capable of conjugating the well-being of individuals and communities with respect for the environment and wise use of resources [4]. Smart cities involve policies and deal with productivity, sustainability, and well-being [5]. They can therefore be ascribed to the realm of utopian models for urban development aimed at providing the context for an ideal and desirable future for humans.

However, several specific factors mark the novelty and peculiarity of the smart city; among others:

- the centrality of digital technologies that provide the nervous system and access/control terminals for a new generation of solutions based on information and algorithms [6–8];
- the lack of a unique model for development: the invention of smart solutions is an open opportunity for innovation and progress where private companies, policymakers, and local administrators, together with single citizens and associations, are agents and stakeholders for smart initiatives to create economic, social, and environmental value [9];
- the global and planetary dimensions of the unprecedented challenges that humanity is facing today.

Smart solutions are made possible by the availability of digital infrastructures and applications. Technologies provide the means for the automation of physical services and facilities, enable the exchange of information and control, and provide access at a distance to facilities and commodities [10]. The scenarios referred to the smart city aim at the rational use of local resources, and at the optimization of efficiency and satisfaction for main services, including mobility, transportation, health care, democratic participation, smart generation, and distribution of energy [7, 10].

Several smart city initiatives are animated by the ambition of creating progress in terms of social and environmental sustainability, but the creation of socio-technical systems based on digital technologies is not necessarily always aligned with the values of sustainable development [11].

The digital transition is associated with enormous economic interests; the creation of digital infrastructures and services has been both demand- and market-driven. The case studies on smart cities reflect different models and ambitions of urban development and include eco-city experiments, systems aimed at improving the efficiency and effectiveness of economic and productive environments, and initiatives for the creation of elitist contexts [7, 9]. Digitalization can be leveraged as a strategic opportunity for industrial and business competition. Big data, artificial intelligence, and automation can increase efficiency and promote the innovation of products, processes, and services; they also support the creation of virtuous synergies between different comparts of industry and services. On the other hand, according to the UN report on Technology and Innovation [12], creating digital facilities and systems requires financial investments and research and technical capabilities. Countries with low capabilities of investment, poor education and research systems (notably for STEM disciplines), and scarce industrial know-how could have severe difficulties in grasping the opportunities offered by digitization. Specific support actions should therefore aim at supporting the digitalization of those nations that have the most significant difficulty in implementing the digital transition. Furthermore, the report refers that "Women are also severely underrepresented in the key area of ICTs, accounting for only 30 percent of total workers in the digital sector in the European Union. Underrepresentation occurs at all levels, but particularly in decision-making positions".

In other words, socially sustainable digitalization requires dedicated education plans, inclusive industrial growth and policies, and the capability to design for diversity.

Altogether, the experiments and the criticism that accompany them do not question the potential of ICTs in the generation of meaningful and desirable solutions for social and environmental progress, but they point out the importance of dealing with the complexity of the impacts and frictions connected to change [13].

In this context, the updating of education programs becomes an important priority to provide the young generations of designers with a suitable background and sense of critics [14–16].

Smart cities are transdisciplinary project fields, and their creation requires the convergence of different domains of knowledge; they are an ideal melting pot where human sciences, design disciplines, and formal sciences collaborate and become contaminated. Smart systems and services, in fact, involve three dimensions and architecture: the physical dimensions, the network of information, and the organization [17]. No development proposal is politically neutral: inevitably, every vision of the future arises from an interpretation of what is right and desirable for individuals, communities, and the environment. Each discipline involved in urban development should develop an ethical reflection on the consequences of project choices and on how to approach the project of the future, considering the impacts and consequences of change in a conscious and responsible way [18].

In the design of smart solutions, the design of services and interaction has the task of understanding people's needs and attitudes and inventing solutions that are meaningful, desirable, and simple. To this end, user-centered design approaches have been developed, which include multiple survey techniques on users and contexts, aimed precisely at collecting insights and identifying design opportunities [19–21].

However, in many cases, the implementation of smart socio-technical systems oriented toward sustainability requires a change of mentality and behavior on the part of people so that they can accept and adhere to new services and systems [22].

Designing smart solutions for sustainable development is not only a question of understanding and responding to people's needs but, rather, of designing to enable changes that are acceptable and beneficial for people and communities.

A question therefore arises: what are the skills that can strengthen the ability of designers to understand human complexity in order to design solutions that garner the support of end users?

This document contributes to answering this question by indicating two specific themes: the theories of Design for Behavior Change as a useful support to the understanding of how to cope with human reluctance for change and the importance of considering the specific gender point of view in order to design services capable of responding more effectively to people's needs.

The remainder of this document is organized as follows: a section is dedicated to illustrating the importance of the active participation of end users in the actions for achieving the objectives of the 2030 Agenda. Two sections are dedicated to the theories of design for behavior change and briefly indicate the results of their application in a university course. Finally, a fourth paragraph focuses on the importance of including the gender perspective in the preliminary research for the design of services for developing inclusive services; it also briefly reports the results of an education experiment.

2. Citizens are the main actors in sustainable development

Humanity today faces multiple challenges on which the future of human beings and the planet will depend. Climate change, dramatic social inequalities, conflicts, pollution, and scarcity of resources ask for actions and responsibility for improving human conditions and conservating natural systems for future generations. Digital technologies play a crucial role in the search for suitable solutions, but technology alone is not sufficient, and every social actor should contribute to the mitigation of the current crises [23, 24].

A meaningful example of the importance of citizens in sustainable development is the energy transition. Energy services are the focus of the seventh goal in UN Agenda 2030 and one of the arenas where smart systems must provide value for people and environments. The targets listed for this goal include the development of technological systems for the production and distribution of clean energy, the substantial increase of renewable sources in the production mix, and the improvement in energy efficiency. Digital systems play a fundamental role in the creation of smart systems that can improve the efficiency and quality of energy services [25] and are essential for managing the integration of local and centralized production. But the energy challenge is not only a matter of production and distribution systems.

The document Climate Change and Energy Renovation Wave by the European Commission for the Environment provides a map of the different opportunities for the specific and local strategies in each country in the European Green Deal: each contest asks for a specific approach to the energy transition. Citizens can be an active part by giving their contributions with behaviors and lifestyles that favor energy savings: they can build or renovate homes and workplaces following sustainability criteria so as to reduce the need for energy in thermal regulation; influence the energy policies of local administrations and the energy industries by asking for transparency on the energy sources negotiating for a fair mix; participate in the co-production of energy, exploiting local resources; contribute to culture change by their advocacy on the value of the energy transition [26].

Figure 1 maps the potential contributions of citizens to the energy transition.

Nowadays, the availability of energy supply under reasonable conditions for consumers and companies – considering quantity, quality, and costs of the services – is not guaranteed in a uniform and fair way in the various parts of the world. According to the UN Agenda, "Health, food security, gender equality, education, economic development and other sustainable development goals critically depend on access to clean, affordable and reliable energy services" [27].

The agenda points out as a primary priority the task of "ensuring universal access to affordable, reliable, and modern energy services, and of expanding infrastructure, and upgrading technology for supplying modern and sustainable energy services for all in developing countries" [28].

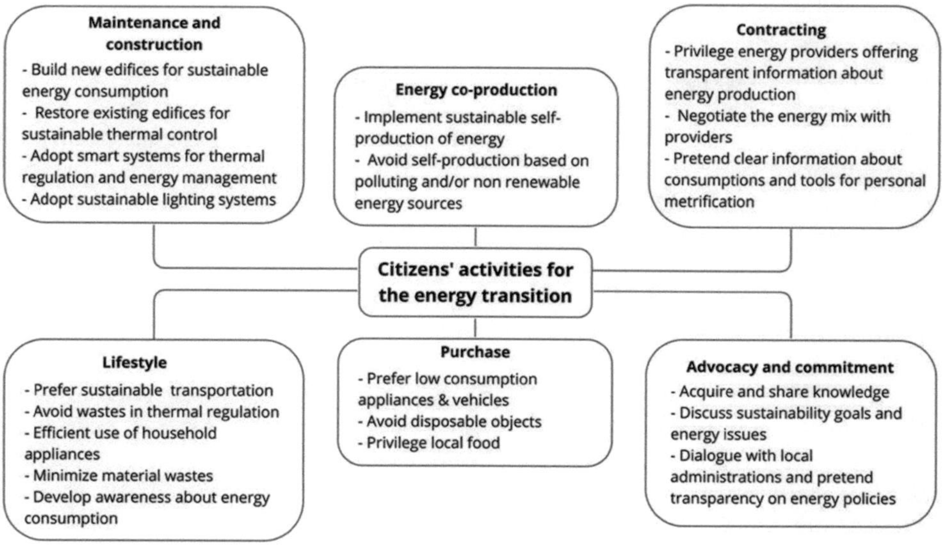

Figure 1.
How citizens can engage themselves in the energy transition.

To produce progress in terms of environmental and social sustainability, the energy transition must therefore involve people so that they collaborate in identifying and implementing the most acceptable, desirable, and effective solutions in local level contexts.

3. Change is a difficult challenge in the design of solutions for sustainable development

The example of the energy transition is emblematic of the fact that to succeed in the challenges of our time, it is necessary to integrate technological skills with the ability to involve citizens as active and fundamental actors of change. In other words, the energy transition requires the convergence and collaboration of different domains of knowledge and expertise, including human and brain sciences, providing insights into human mental models and processes and the capability to develop languages, experiences, and storytelling to favor awareness and engagement. This is necessary for most changes required by the sustainability goals. According to Linner et al. [29], the agenda goals require social research for understanding how the transformation takes place in local contexts and how it is associated with changes in sense-making and social values.

The change of mindsets and behaviors is not an easy task. The lack of knowledge, suitable means, and technical skills can hinder the active participation of citizens in change. But often, the implementation of new socio-technical systems also encounters obstacles related to the lack of understanding of the importance of the transformations required by the sustainable development goals and to scarce motivation for personal engagement. The change of mindset is not straightforward even in front to sound scientific data demonstrating evidence [30–33]. According to these authors, the reluctance toward a change of opinion is very strong in individuals, regardless of their education and cognitive capabilities. From the point of view of social and personal stability, attachment to the personal convictions and vision of reality can be useful as it contributes to sustaining determination and perseverance even in the face of crises and difficulties.

In the implementation of systems for sustainable development, reluctance to change is a factor of complexity to cope with, also demanding specific attention and dedicated knowledge. For this purpose, the theories of Design for Behavior Change (DfBC) can offer a valuable contribution.

4. Including design for behavior change in the conceptual toolbox for the design of digital services and systems

Design for Behavior Change (DfBC) is a set of design theories based on the application of behavioral sciences and aimed at supporting the project of solutions requiring a change of attitude, behavior, or mindset in users. According to Niedderer et al. [34], "Design for behavior change is concerned with how design can shape or influence human behavior and sustainable innovation".

While it can be argued that most design work is addressed to have an impact on users, according to Lockton [35], "Systems intentionally designed to influence behavior different from that usually associated with the situation or in situations where a user would not otherwise have a strong idea of what to do (e.g., with an unfamiliar interface), represent a degree of designer intent beyond this".

It is important to point out that DfBC is not aimed at the persuasion of the users; it aims instead to understand what are the factors that prevent people from adopting a certain behavior even when it is associated with obvious individual and collective advantages. Application of DfBC could impact solutions to support health-friendly lifestyles; could empower people to reduce the use of energy and water resources, food waste, and correct waste management; could give a contribution to enabling responsibility toward the common goods and the territory; could promote the adoption of collaborative behaviors; could encourage responsible behaviors in mobility. According to Wendel [36], DfBC assumes that the human mind has limits in attention and willpower and that human activities depend on both conscious and unconscious thinking, with habits and automatism often governing our behaviors while our decision processes are influenced by the context. These assumptions are coherent with the findings of the scientists rewarded with the Nobel prize to the psychologist Daniel Kahneman and the economist Richard H. Thaler.

Figure 2 shows the main obstacles that can hinder people from changing behavior even when they understand the benefits of change.

Other authors instead frame the obstacles to behavior change in terms of physical vs. phycological capabilities, reflective vs. automatic motivation, and physical vs. social opportunities.

Withe et al. [37] produced a review on strategies for shifting consumer behaviors to be more sustainable. Their research includes several factors such as social influence, habits, sense of self, feelings, and cognitions. The authors propose drivers for positive engagement of citizens and consumers that address the importance of operating on symbolic attributes of innovative proposals. They remark the focus on broadening the sense of self associated with the adoption of sustainable lifestyles and preferences; the focus on positive emotions, empathy, and moral elevation as leverages in sustainable change.

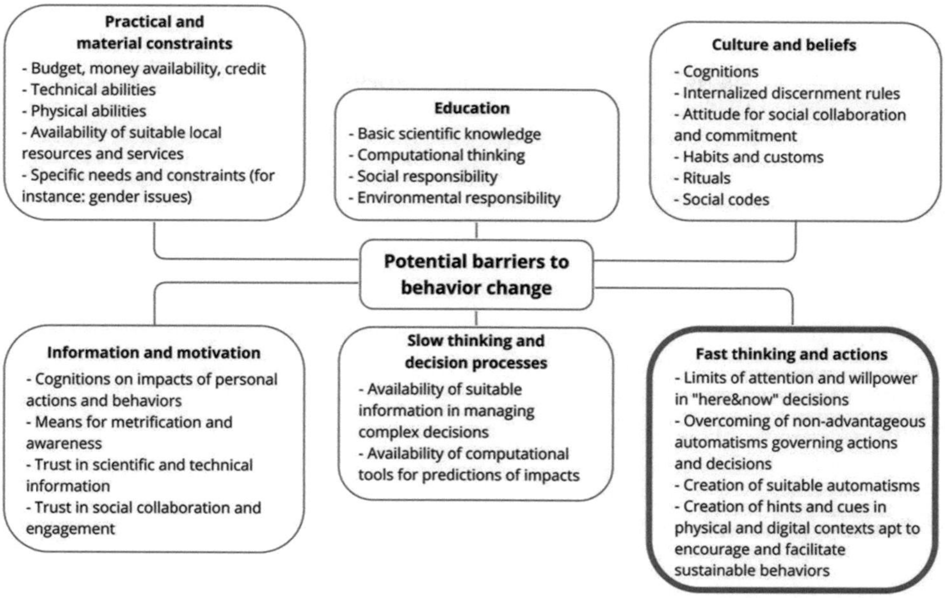

Figure 2.
The several factors that can hamper the availability of citizens for behavior change. The bold frame points the attention to the specific focus of theories for DfBC.

Michie et al. present a systematic review on the effectiveness of different techniques for behavior change in health care [38]. Despite the available data documenting the effectiveness of the techniques evaluated by the authors, the paper points out the opportunities and needs for systematic and comparative studies on the topic that, with the support of data collection and analysis, could provide new paradigms for research in health care.

Altogether, academic literature on DfBC shows its potential as a conceptual aid in the design of services requiring change of behavior and/or mindset in final users and stakeholders.

The DfBC theories can be useful in the design of digital solutions for sustainability? To answer this question, during the academic year 2021-2022, an educational experiment was conducted with the students of the MsC in Digital and Interaction Design at the Design School of Politecnico di Milano within the course of UX Design. The course aims at teaching how to design digital services and applications, and it includes preliminary research on contexts, stakeholders, and users to collect insight and design opportunities. The class counted approximately 60 students from 14 different countries, working in 9 teams. Students attended lectures on the sustainability goals of the 2030 agenda; the lessons on theories and case studies on DfBC introduced students to the difficulties of inducing a change in behavior and provided the conceptual tools to frame the bottlenecks that can prevent it.

Although initially, a few students declared difficulties in understanding the theories of the DfBC, during the course, the teams progressively recognized its value. Behavioral and cognitive sciences provide new models to describe the understanding of the complexity of the information processing and decision-making processes that govern human behavior; the acknowledgment of this knowledge and its implications may require a change of mindset in designers that is not easy to accept at first.

The teams were free to look for design opportunities, scouting for innovative forms to produce digital or digital/physical services; they were asked to conduct preliminary research, devise concepts, and prototype the applications for final assessment with stakeholders and users. The concepts proposed by the students included a service to reduce food waste and a system to encourage the consumption of seasonal food, applications to counter impulsive purchases in the field of clothing, a physical/digital system for education in the conscious use of mobile phones for children, a service for reducing the use of disposable items in the hotel sector, an application for self-control and awareness in interpersonal communication on digital channels.

The results of the course showed that the theories of the DfBC can offer a contribution to university education for the digital services project capable of making students expand their user-centered culture and develop a complexity-aware attitude. According to the students, the discussion on the DfBC theories also stimulated their thinking about the ethical issues in the design of digital services.

5. Exploiting human diversity in the design for sustainable development

The scientific knowledge that describes the general human reluctance toward change can be valuable for the design of effective systems involving people in changes for sustainability.

This topic is twined with another one that should be considered fundamental in university-level training education for the inclusive design of smart services and systems, i.e., the attention to human diversity.

The issue of human diversity has been vastly explored in research and theories of service and interaction design: as stated by D. Evans [39], in service and interaction design, it is important to start from the assumption "that one size doesn't fit all" and that innovation asks for the capability of coping with human differences. Designers can refer to literature from different disciplines investigating human diversity to extract knowledge to design for inclusivity. Examples of this are the research on human disposition and the impacts of age on the relationship with technologies [40, 41]. On the other hand, interaction and service design rely on multiple methodologies for the human-centered design that have been developed for research on users and contexts. Over the last half-century, the methodologies for user-centered design kept on evolving, together with the conception of the role of the final user in the design process: from being the recipient of the project to being a stakeholder in the invention and implementation of services. In a recent paper, Auernhammer et al. [42] reported research outcomes on the evolution of Human Centred Design over the last half-century. The document also presents a map visualizing the various approaches adopted in investigating user needs for the different design purposes and branches: from Ergonomics to the studies on human-computer interaction and design for all. The document enlists the fields of application for the studies on users, ranging from interaction design to universal design and design for inclusion. It also traces the main steps of the evolution: from ergonomics to participatory design and the growing importance of psychological theories.

Despite the wide deployment of methods and approaches for research in users, research in this sector has paid little attention to the specific needs and viewpoints of women. For example, Beebeejaun [43] remarks that gender issues are yet scarcely considered in the academic literature, despite the fact that they should be a pivotal focus in considering social diversity.

Gender equality is the focus of the fifth goal on the UN Agenda, and this specific focus is important and a priority not only for women but for all social communities since, as clearly indicated in the UN Agenda, "Women are not only the hardest hit by this pandemic, they are also the backbone of recovery in communities. Putting women and girls at the center of economies will fundamentally drive better and more sustainable development outcomes for all, support more rapid recovery, and place the world back on the footing to achieve the Sustainability Development goals" [44, 45].

The tasks in the fifth goal focus on the mitigation of gender-based violence, the need for dedicated social protection and economic stimuli, and the importance of inclusion of women and girls in planning and decision-making. Asteria [46] points out that "Smart cities must develop gender sensitivity and awareness about the needs of women by using inclusion mechanisms."

Furthermore, the goal points out the importance of adopting gender perspectives in data and coordination mechanisms that are crucial for the future of smart cities. Special attention should be paid, in fact, to bias in AI algorithms due to uneven availability of data or to gender blindness in data gathering. Bias in data gathering and processing impacts not only women but also on fairness and the capability to manage diversity in the whole society [47, 48].

The adoption of a gender perspective is valuable also in the design or optimization of services and systems dedicated to all citizens regardless of gender, such, for example, in the case of mobility and transport services. The studies on this topic that consider a gender perspective show the specific needs, priorities, and requirements of women. According to Chang [49], the perception of spaces, including affordances and risks, is gender-dependent. The World Development Report 2012 [50] reports specific women's needs related to their activities, their role in their families, and activities as caregivers. Women have specific needs regarding the time schedule, the physical and ergonomic characteristics of the vehicles, and the personal safety of vehicles and stations. In several countries, women have limited access to transportation due to their limited economic power and technical skills and reduced access to ownership of transportation means. Limited access to transportation and mobility has an impact on opportunities for education, work, health care, culture, and leisure. Some countries, such as Sweden and Germany, have developed gender-oriented services and initiatives to mitigate the gap in opportunities for women and facilitate the adhesion to sustainable services [51, 52].

The digression on the issue of mobility and transport is just one of the examples that can be cited to illustrate how considering the point of view of women on a specific service or system leads to the understanding of specific factors that can jeopardize access and acceptability for a large part of the population. The same reasoning could be extended to sectors such as smart working systems, medical care, and others.

In the academic year 2020-2021, during the period of social restrictions due to the pandemic due to Covid-19, an educational experiment was conducted to investigate the potential of adopting a gender point of view within the process. Design for digital services and systems. The experiment was conducted with the students of the MSc in Digital and Interaction Design at the School of Design of the Politecnico di Milano and involved about 60 students who conducted preliminary research and development of innovative digital service concepts. The course was conducted with the participation of the Consulta Femminile Interassociativa di Milano (Consfim, https://www.consultafemminilemi.com) and the local club Soroptimist International Milano alla Scala. Among other activities, these female associations are promoting the debate on smart cities with a gender-focused approach aimed at orienting the local administration toward inclusive policies for development. Similarly as in the education experiment on DfBC reported in the first part of this document, students were asked to generate concepts of digital services with a double-diamond design approach, freely scouting for design opportunities through research on users and context. However, in this course, students were asked to investigate the specific point of view of women and collect specific design hints related to the diversity of gender. The students expressed great appreciation for this experience, and several reported that the analysis of the perspective of gender enabled a deeper and more articulated understanding of needs and opportunities to create value for the users. The concepts developed by the teams included mobility and transportation (focus on access to sharing-service, safety in public transportation and urban mobility, last-mile mobility), social life for elderly people, mental health, family activities for children's care and education, and the exploitation and care of public spaces. The female association that patronized the experiment expressed high appreciation for the outcomes of preliminary research and concept development; they consider the contents as valuable to inspire and orient the dialog with the local public administration for gender-sensible development policies. Further information can be found in a dedicated paper [53].

6. Conclusions

According to the idea of a smart city, urban centers are conceived as dynamic entities capable of fast evolutions and adaptation far beyond the constraints of physical structures that define the city layout and material configuration. Implementing the new socio-technical systems should give solutions for the challenges of our time if they are framed in the sustainable development goals of the UN Agenda 2030. The digital infrastructures can provide data enabling efficient use of resources and services by offering the means for understanding and answering the diversified needs of citizens. The conjugation of the principles of environmental and social sustainability with the development models of smart cities is fertile. It can offer significant opportunities for innovation and progress to benefit people and the environment. As clearly indicated by the UN documents, the exploitation of this potential asks for large and diversified involvement and engagement of citizens.

This document reports research aimed at developing education at the university level for designing interactive solutions and digital services coherent with the principles of social and environmental sustainability. The research starts from the assumption that the capability to deal with the complexity of social systems and cope with human diversity is a requirement for the design of inclusive and desirable development.

The first section of the document presents the research background; it points out the importance of confronting the innovation paradigms of smart cities with the development goals of the UN Agenda 2030 and of updating the contents for design education considering the outcomes of behavioral sciences that can be useful in the project of inclusive solutions.

Section two refers to the energy transition program as a case study showing the diversified modality in which the citizens can contribute to the implementation of the goals for sustainable development. The third part reports literature references about the complexity of producing changes in behaviors and mindsets.

The deployment of the main theories on design for behavior change is the main content of section four. This knowledge enriches the theoretical and methodological competencies for designing effective services and physical-digital systems; it provides suitable content for university design education, as demonstrated by a teaching experience performed at the Design School of Politecnico di Milano, also reported in the chapter.

The fifth and last part of the document focuses on gender issues and on the importance of considering women's perspectives in the design of inclusive and effective socio-technical systems. Gender equality is a primary requirement for democracy, and it is a specific goal of the UN Agenda. The chapter points out the value of considering gender issues in the development of general-purpose services, and it reports the case study of transportation as a meaningful example. Academic literature shows that gender-oriented studies are currently very scarce; to ensure inclusive development, it is, therefore, necessary to develop dedicated research. The document reports the results of an educational experiment for service design with a gender perspective, which demonstrate the effectiveness of this theme in university education.

To summarize, this chapter is an act of advocacy for the inclusion of two topics in design education. Cognitive psychology and behavioral sciences provide helpful knowledge for education on inclusive design. The focus on the gender perspective is also presented as a valuable source of inspiration in the development of innovative socio-technical systems. Both topics are complex and require investigations and discussions, but experiments demonstrated their suitability for university education and the cultural growth of young designers.

Future developments of the research will focus on both, with a special focus on how to create motivation and engagement for sustainability.

The document "New threats to human security in the Anthropocene" [54] points out the importance of working for improving the availability of solidarity, agency, and engagement. These goals ask for multidisciplinary research and experimentation and more dialog between the social stakeholders of development.

Acknowledgements

The author thanks the president of the Women's Council of Milan Laura Caradonna for her generous collaboration and the conversations on sustainable development issues in a gender proposal. She also thanks the teaching assistants Laura Varisco; Alessandra Mazzola and Martina Marzola, who collaborated in the classroom activities.

Author details

Margherita Pillan
Dipartimento del Design, Politecnico di Milano, Milano, Italia

*Address all correspondence to: margherita.pillan@polimi.it

References

[1] Angelidou M. Smart cities: A conjuncture of four forces. Cities. 2015. ISSN 0264-2751;**47**:95-106. DOI: 10.1016/j.cities.2015.05.004

[2] Gropius W. Letter to Thomas Maldonado. Ulm, Zetschrift der Hochschule für Gestlatung. N.10-11, 1964 in Fiedler J Bauhaus 2006 Tandem Verlag GmbH. Giabudo/Könemann Editor

[3] Manzini E. Design, When Everybody Designs. An Introduction to Design for Social Innovation. Cambridge, Massachusetts: The MIT Press; 2015

[4] Yigitcanlar T, Md K, Buys L, Ioppolo G, Sabatini-Marques J, Moreira da Costa E, et al. Understanding 'smart cities': Intertwining development drivers with desired outcomes in a multidimensional framework. Cities. 2018, ISSN 0264-2751;**81**:1-16. DOI: 10.1016/j.cities.2018.04.003

[5] Mora L, Deakin M. Chapter 1 - Moving beyond the Smart City Utopia. Untangling Smart Cities. Elsevier; 2019. pp. 1-17, ISBN 9780128154779. DOI: 10.1016/B978-0-12-815477-9.00001-3

[6] Yigitcanlar T, Kankanamge N, Vella K. How Are SmartCity Concepts and Technologies Perceived and Utilized? A Systematic Geo-Twitter Analysis of Smart Cities in Australia. Journal of Urban Technology; 2020;**28**(1-2):135-154. DOI: 10.1080/10630732.2020.1753483

[7] Arroub A, Zahi B, Sabir E, Sadik M. A literature review on smart cities: Paradigms, opportunities and open problems. Proc. International Conference on Wireless Networks and Mobile Communications (WINCOM); 2016. p. 180-186. DOI: 10.1109/ WINCOM.2016. 7777211

[8] Camero A, Alba E. Smart City and information technology: A review. Cities. 2019, ISSN 0264-2751;**93**:84-94. DOI: 10.1016/j.cities.2019.04.014

[9] Trindade EP, Hinnig MPF, da Costa EM, et al. Sustainable development of smart cities: A systematic review of the literature. Journal of Open Innovation. 2017;**3**:11. DOI: 10.1186/ s40852-017-0063-2

[10] Lee J, Babcock J, Pham TS, Kang M. Smart city as a social transition towards inclusive development through technology: A tale of four smart cities. Internationl Journal of Urban Sciences. 2020. p. 1-27. DOI: 10.1080/12265934.2022.2074076

[11] Ahvenniemi H, Huovila A, Pinto-Seppä I, Airaksinen M. What are the differences between sustainable and smart cities? Cities. 2017 ISSN 0264-2751;**60**(Part A):234-245. DOI: 10.1016/j. cities.2016.09.009

[12] United Nation Conference on Trade and Development. Technology and Innovation Report 2018. Switzerland: United Nations Publication; 2018 UNCTAD/TIR/2018. ISBN 978-92-1-112925-0 e-ISBN 978-92-1-363310-6

[13] Mazza P. Education & Smart Cities: The role of the goals of agenda 2030 for sustainable development of smart cities. International Journal of Innovative Studies in Sociology and Humanities. Open Access. 2021;**6**(2):24-31. DOI: 10.20431/2456-4931.0602003

[14] Alcantud-Díaz M. Research, Teaching and Actions in Higher Education on the UN Sustainable Development Goals. Cambridge, UK: Cambridge Scholar Publishing; 2021

[15] Fia M, Ghasemzadeh K, Paletta A. How Higher Education Institutions Walk Their Talk on the 2030 Agenda: A Systematic Literature Review. High Educ Policy; Journal of Higher Education Policy. Springer. 2022. DOI: 10.1057/s41307-022-00277-x

[16] Zaleniene I, Pereira P. Higher education for sustainability: A global perspective. Journal of Geography and Sustainability. Vol. 2. p 99-1062021. DOI: 10.1016/j.geosus.2021.05.001

[17] Pavlovic M, Bier H, Pillan M. Ambient UX for cyber-physical spaces. Actuated and performative architecture emerging forms of human-machine interaction. SPOOL. 2020;7(8):27-36. DOI: 10.7480/spool.2020.3

[18] Helbing D, Fanitabasi F, Giannotti F, et al. Ethics of smart cities: Towards value-sensitive design and Co-Evolving City life. Sustainability. 2021;**13**(20):1-25. DOI: 10.3390/su132011162

[19] Benion D. Designing User Experience. London, England: Pearson Education Limited; 2019

[20] Stickdorn M, Schneider J. This Is Service Design Thinking. Amsterdam, The Nederlands: John Wiley & Sons; 2011

[21] Cooper A, Reimann R. About Face 2.0: The Essentials of Interaction Design. Amsterdam, The Nederlands: John Wiley & Sons; 2003 ISBN 0764526413

[22] Radziejowska A, Sobotka B. Analysis of the social aspect of smart cities development for the example of smart sustainable buildings. Energies. 2021;**14**:4330. DOI: 10.3390/en14144330

[23] Cathelat B et al. Smart - Cities Shaping the Society of 2030. United Nations Educational, Scientific and Cultural Organization (UNESCO); 2019

[24] Thinyane M. Engaging Citizens for Sustainable Development. Macau, China: A Data Perspective. Unu-cs; 2018

[25] United Nations General Assembly. Resolution Adopted by the General Assembly on 25 September 2015. Distr.: General; 2015

[26] Wahlund M, Palm J. The role of energy democracy and energy citizenship for participatory energy transitions: A comprehensive review. Energy Research & Social Science. 2022, ISSN 2214-6296;**87**:1-19. DOI: 10.1016/j.erss.2021.102482

[27] UN Sustainable development goals, Goal 7. https://www.un.org/sustainabledevelopment/energy/

[28] UN Sustainable development goals, Goal 7. Ensure Access to Affordable, Reliable, Sustainable and Modern Energy for All. https://unstats.un.org/sdgs/report/2017/goal-07/

[29] Linnér BO, Wibek V. Sustainability Transformation – Agents and Drives across Societies. Earth System Governance. Sweden: Cambridge University Press; 2019

[30] Gesiarz F, Cahill D. Sharot T evidence accumulation is biased by motivation: A computational account. PLoS Computational Biology. 2019;**15**(6):e1007089. DOI: 10.1371/journal.pcbi.1007089

[31] Kappes A, Harvey LAH, et al. Confirmation bias in the utilization of others' opinion strength. Nature Neuroscience. 2020;**23**:130-137. DOI: 10.1038/s41593-019-0549-2

[32] Sharot T, Sunstein CR. How people decide what they want to know. Nature Human Behaviour. 2020;**4**:1-6. DOI: 10.1038/s41562-019-0793-1

[33] West R, Michie SA. Brief introduction to the COM-B model of behavior and the PRIME theory of motivation. Qeios Journal2020. p. 1-6. DOI: 10.32388/WW04E6.2

[34] Niedderer K et al. Design for Behaviour Change as a Driver for Sustainable Innovation: Challenges and Opportunities for Implementation in the Private and Public Sectors. International Journal of Design. 2016;**10**(2):67-85

[35] Lockton D, Harrison DJ, Stanton NA. The design with intent method: A design tool for influencing user behaviour. Applied Ergonomics. 2010;**41**(3):382-393. DOI: 10.1016/j.apergo.2009.09.001

[36] Wendel S. Designing for Behavior Change. US: O'Reilly Media Inc; 2013 ISBN: 9781449367626

[37] White K, Habib R, Hardisty DJ. How to SHIFT consumer behaviors to be more sustainable: A literature review and guiding framework. Journal of Marketing. 2019;**83**(3):22-49. DOI: 10.1177/0022242919825649

[38] Atkins L, Michie S. Designing interventions to change eating behaviours. Proceedings of the Nutrition Society. 2015;**74**(2):164-170. DOI: 10.1017/S0029665115000075

[39] Evans D. Bottlenecks: Aligning UX Design with User Psychology. New York, US: Apress; 2017. DOI: 10.1007/978-1-4842-2580-6 ISBN: 978-1-4842-2579-0

[40] McCrae RR, Costa RR Jr. A five-factor theory of personality. Handbook of personality. Theory and Research. 1999;**2**:139-143. Guilford

[41] Fleming A, Mason C, Paxton G. Discourses of technology, ageing and participation. Palgrave Communication. 2018;**4**:54. DOI: 10.1057/s41599-018-0107-7

[42] Auernhammer J, Zallio M, Domingo L, Leifer LJ. Facets of human-centered design: The evolution of designing by, with, and for Peoplo. Preprint. 2022 https://www.researchgate.net/publication/355796639

[43] Beebeejaun Y. Gender, urban space, and the right to everyday life. Journal of Urban Affairs. 2017;**39**:3. DOI: 10.1080/07352166.2016.1255526

[44] https://www.un.org/sustainable development/gender-equality/

[45] Global Gender Gap Report. World Economic Forum. Geneva, Switzerland: Published World Economic Forum, Cologny; 2020;**2019**. ISBN-13: 978-2-940631-03-2

[46] Asteria D, Jap JJK, Utari D. A gender-responsive approach: Social innovation for the sustainable Smart City in Indonesia and beyond. Journal of International Women's Studies. 2017;**21**(6):193-207. Available from: https://vc.bridgew.edu/jiws/vol21/iss6/12

[47] Nadeem A, Abedin B, Marjanovic O. Gender bias in AI: A review of contributing factors and mitigating strategies. ACIS 2020 Proceedings. 2020;**27**:259-270. Available from: https://aisel.aisnet.org/acis2020/27

[48] Leavy S. Gender bias in artificial intelligence: The need for diversity and gender theory in machine learning. Proceedings of theIEEE/ACM 1st International Workshop on Gender Equality in Software Engineering, Gothentburg Sweden. New York, US: Conference Publishing Services; 2018. p. 14-16. ISBN: 978-1-4503-5738-8

[49] Chang J, Choi J, An H, Chung H. Gendering the smart city: A case study of Sejong City, Korea. Cities. 2022, ISSN 0264-2751;**120**:1-11. DOI: 10.1016/j.cities.2021.103422

[50] World Bank. World Development Report 2012: Gender Equality and Development. 2012 https://openknowledge.worldbank.org/handle/10986/4391

[51] https://womenmobilize.org/

[52] Swedish Government Official Reports. Gender Equality in Public Services. A Book of Ideas for Managers and Strategists from the Swedish Gender. Sweden: Mainstreaming Support Committee; 2007

[53] Pillan M, Marzola M. Experimenting the Role of UX Design in the Definition of Gender-Sensitive Service Design Policies. Bilbao: DRS2022; 2022. DOI: 10.21606/drs.2022.495

[54] United Nations Development Programme. New Threats to Human Security in the Anthropocene Demanding Greater Solidarity. New York, USA: Published by the United Nations Dev Progr; 2022

Chapter 3

Analysis of Solution Diversity in Topic Models for Smart City Applications

Toshio Uchiyama and Tsukasa Hokimoto

Abstract

Topic models are known to be useful tools for modeling and analyzing high-dimensional count data such as documents. In a smart city, it is important to collect and analyze citizens' voices to discover their concerns and issues. Topic modeling is effective for the above analysis because it can extract topics from a collection of documents. However, when estimating parameters (solutions) in topic models, various solutions are reached due to differences in algorithms and initial values. In order to select a solution suitable for the purpose from among the various solutions, it is necessary to know what kind of solutions exist. This chapter introduces methods for analyzing diverse solutions and obtaining an overall picture of the solutions.

Keywords: topic model, diversity of solution, normalized mutual information, typification of solutions, topic distribution, word distribution, information-theoretic clustering

1. Introduction

Probabilistic latent semantic analysis (PLSA: probabilistic latent semantic analysis) [1] and latent Dirichlet analysis (LDA: latent Dirichlet allocation) [2] are known as topic models to analyze count data such as documents (text data). In a smart city, it is important to collect and analyze citizens' voices to discover their concerns and issues. Topic modeling is effective for the above analysis because it can extract topics from a collection of documents. However, when estimating parameters (solutions) in topic models, various solutions are reached due to differences in algorithms and initial values. There could exist a lot of local optimal solutions that are distinct but are equally optimized in the objective function (**Figure 1**). Since each of these solutions presents an interpretation of data, they are meaningful and worth using. In order to select a solution suitable for the purpose from among the various solutions, it is necessary to know what kind of solutions exist. This chapter introduces methods for analyzing diverse solutions and obtaining an overall picture of the solutions.

The solution, which is the set of parameters estimated in topic models, has a topic distribution $\boldsymbol{\theta}$ and a word distribution $\boldsymbol{\phi}$. A topic distribution $\boldsymbol{\theta}$ is tied to each

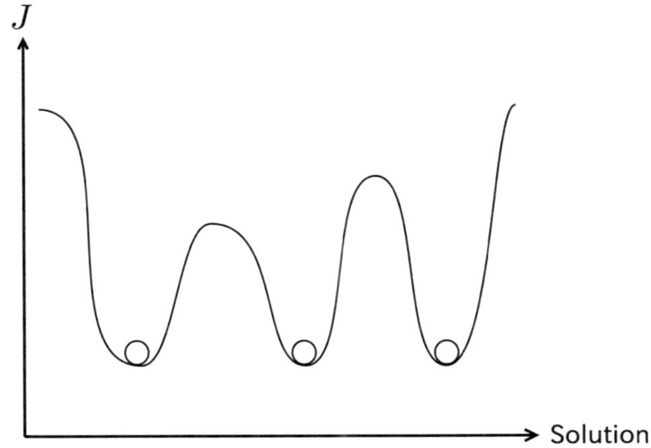

Figure 1.
An objective function for topic model J and local optimal solutions.

document, and it represents the mixture ratio of topics in the document. The word distribution $\boldsymbol{\phi}$ represents the probability of occurrence of each word in a topic. The methods presented here analyze diverse solutions using topic and word distributions, respectively.

The method [3, 4] using topic distribution $\boldsymbol{\theta}$ defines the normalized mutual information (NMI) [5] that can be calculated for two solutions as the similarity between them, and assign coordinate values to them by multidimensional scaling (MDS) [6]. The coordinate values can be used to visualize the distribution of solutions in low-dimensional space.

Word distribution $\boldsymbol{\phi}$ directly represents topic characteristics and is easy for humans to understand, making their analysis valuable. Specifically, clustering and network representation of similar relations are used to obtain groups of word distributions that are similar to each other. Each solution is then typified based on the frequency distribution of the groups. As a result, several typical solutions and word distributions that could be taken were successfully represented in a human-understandable form [7].

The related studies are shown below. As for analyzing multiple solutions, there are studies on clustering. In these studies, the Rand index or NMI is used to define the distance or similarity between solutions. Then, a non-redundant alternative solution for a given solution is found [8], several non-redundant are searched [9], and solutions are visualized by dendrogram [10]. This chapter focuses on topic model which includes hard clustering as a special case. The article [11] is a study of visualizing the solution of a topic model, but for a single solution. Few studies analyze and visualize multiple solutions in topic models.

The experiments deal mainly with text data of news articles and show the distribution of the solutions and the typical topics in the topic model. We expect that the proposed methods will contribute to the discovery of problems in smart cities.

2. Topic models and estimation of its solution

Topic models could be described by a generative probability model as shown in **Figure 2**. Shaded circles represent observed variables and white circles represent

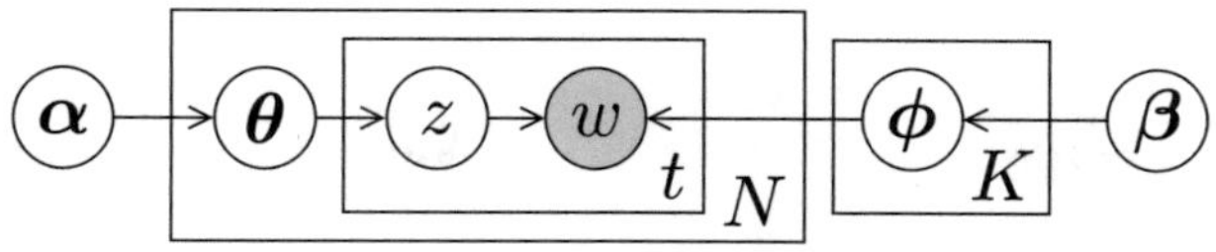

Figure 2.
Graphical model for topic model.

unobserved variables. The square plates represent repetitions, and the number in the lower right corner indicates the number of repetitions. Items inside the plate are conditionally independent given the variable outside the plate. In fact, the observed variables are "words" and the observed data, which is the accumulation of these words, is a document.

Assume that there are topics $k(k = 1, \dots, K)$, document $i(i = 1, \dots, N)$, and M types of words in the total of documents. Also, assume that each topic k has a word distribution $\boldsymbol{\phi}^k = \{\phi_1^k, \dots, \phi_m^k, \dots, \phi_M^k\}$ and the document has a topic distribution $\boldsymbol{\theta}^i = \{\theta_1^i, \dots, \theta_k^i, \dots, \theta_K^i\}$. For each document i, the following is repeated t^i times. A topic z is assigned based on the topic distribution $\boldsymbol{\theta}^i$, and a specific word w is generated based on the word distribution $\boldsymbol{\phi}^z$. All words generated for document i are aggregated by type and made into an observed value vector $\boldsymbol{x}^i = w_1^i, \dots, w_M^i$, where w_m^i denotes the frequency of m-th word. Hence, l^1-norm of i-th observed value vector $\sum_m |x_m^i|$ equals t^i. All observed value vectors are denoted by $\mathcal{X} = \{\boldsymbol{x}^1, \dots, \boldsymbol{x}^N\}$, which is the vector representation of all documents. $\boldsymbol{\alpha}$ and $\boldsymbol{\beta}$ are hyperparameters for the prior probability distributions of the topic distribution $\boldsymbol{\theta}$ and the word distribution $\boldsymbol{\phi}$ respectively, and a uniform Dirichlet distribution is assumed here.

Given the probability $P(m|i)$ that word m is generated in document i, $P(k|i)P(m|k) = \theta_k^i \phi_m^k$ if the k-th topic is assigned, so $P(m|i) = \sum_{k=1}^K \theta_k^i \phi_m^k$ considering all topics. The number of words m in document i equals x_m^i, and all word generations are independent. Therefore, the simultaneous probability for all document generations can be represented by a multinomial distribution

$$\prod_{i=1}^{N} A^i \prod_{m=1}^{M} \left(\sum_{k=1}^{K} \theta_k^i \phi_m^k \right)^{x_m^i}, \tag{1}$$

where A^i is the number of combination for document i. Taking the logarithm of Eq. (1) yields

$$\sum_{i=1}^{N} \log A^i + \sum_{i=1}^{N} \sum_{m=1}^{M} x_m^i \log \left(\sum_{k=1}^{K} \theta_k^i \phi_m^k \right). \tag{2}$$

The first term is a constant for the observed value vectors $\mathcal{X}$, and the second term is the objective function to be maximized, when inferring parameters. A set of parameters $\{\boldsymbol{\theta}, \boldsymbol{\phi}\}$ of this function is a solution to be inferred. In the experiment, we use the equivalent perplexity:

$$\exp \left(\frac{-\sum_i^N \sum_m^M x_m^i \log \left(\sum_k^K \theta_k^i \phi_m^k \right)}{\sum_i^N \sum_m^M x_m^i} \right). \tag{3}$$

There are various algorithms for estimating the model parameters, including collapsed Gibbs sampling (CGS) [12], variational Bayesian estimation (VB) [2], collapsed variational Bayesian estimation (CVB) [13], maximum likelihood estimation (ML) [1], maximum a posteriori probability (MAP) estimation [14]. From these, we use MAP estimation, CGS as a sampling approximation method, and CVB0 [15], which uses zero-order approximation of CVB, as a variational approximation method to estimate diverse solutions.

The update formulas used in the iterations in MAP estimation of $\boldsymbol{\theta}, \boldsymbol{\phi}$ are

$$\hat{\phi}_m^k = \frac{\eta_m^k + \beta - 1}{\sum_{m'} \eta_{m'}^k + M(\beta - 1)}, \quad \hat{\theta}_k^i = \frac{\eta_k^i + \alpha - 1}{\sum_{k'} \eta_{k'}^i + K(\alpha - 1)}, \tag{4}$$

where η_m^k is the number of occurrence about the word of type m at topic k and η_k^i is the number of assignment of topic k to data i. These can be calculated by

$$\rho_{imk} = \frac{\theta_k^i \phi_m^k}{\sum_{k'} \theta_{k'}^i \phi_m^{k'}}, \quad \eta_m^k = \sum_i x_m^i \rho_{imk}, \quad \eta_k^i = \sum_m x_m^i \rho_{imk}. \tag{5}$$

The updates at each iteration in CGS estimation can be written as

$$\hat{\phi}_m^k = \frac{\eta_m^k + \beta}{\sum_{m'} \eta_{m'}^k + M\beta}, \quad \hat{\theta}_k^i = \frac{\eta_k^i + \alpha}{\sum_{k'} \eta_{k'}^i + K\alpha}, \tag{6}$$

using the sampled topic set. When sampling for the j-th word w_{ij} in data i, the probability that the topic is k when the type of this word is m follows

$$P(z_{ij} = k | \mathbf{Z}_{\backslash ij}, \mathcal{X}) \propto \frac{\eta_{m \backslash ij}^k + \beta}{\sum_{m'} \eta_{m' \backslash ij}^k + M\beta} \left(\eta_{k \backslash ij}^i + \alpha \right), \tag{7}$$

where $\backslash ij$ denotes that the information about the word w_{ij} under focus is excluded, and $\mathbf{Z}_{\backslash ij}$ denotes the topic set excluding the topic z_{ij} of the word w_{ij}.

In CVB0 estimation, for the j-th word w_{ij} in data i, the burden rate that the topic is k when the type of this word is m follows

$$\rho_{ijk} \propto \frac{\eta_{m \backslash ij}^k + \beta}{\sum_{m'} \eta_{m' \backslash ij}^k + M\beta} \left(\eta_{k \backslash ij}^i + \alpha \right), \quad \sum_k \rho_{ijk} = 1. \tag{8}$$

The expectations of η_m^k and η_k^i are respectively estimated by $E[\eta_m^k] = \sum_{i,j | w_{ij} = m} \rho_{ijk}$ and $E[\eta_k^i] = \sum_j \rho_{ijk}$. Estimation of ϕ_m^k and θ_k^i are obtained by

$$\hat{\phi}_m^k = \frac{E[\eta_m^k] + \beta}{\sum_{m'} E[\eta_{m'}^k] + M\beta}, \quad \hat{\theta}_k^i = \frac{E[\eta_k^i] + \alpha}{\sum_{k'} E\left[\eta_{k'}^i\right] + K\alpha}. \tag{9}$$

Estimation in CVB0 proceeds by alternating between estimating ϕ_m^k and θ_k^i and updating the burden rate ρ_{ijk} in Eq. (8) [15].

Initial value setting using information-theoretic clustering is applicable in MAP estimation. It was shown that weighted information-theoretic clustering is a special case of topic models (see Appendix A), and it was confirmed that using the clustering results for initial value setting yields a better solution than using random initial value setting [16]. Specifically, the method is to smooth the word distribution obtained from clustering by adding a small value and use it as the initial word distribution $\boldsymbol{\phi}$.

3. An analysis method using topic distribution

This section presents the method [3, 4] to assign coordinate values to solutions using the topic distribution $\boldsymbol{\theta}$. This allows visualization of solutions.

The Normalized Mutual Information (NMI) is known as external criterion for evaluation of clustering [5], but can be applied to solutions in topic models as follows.

There are two solution, A and B for the same set of documents. A has J topics $A^j (j = 1, \dots, J)$ and B has K topics $B^k (k = 1, \dots, K)$. The topic distribution of document $i (i = 1, \dots, N)$ for A and B are denoted by θ^i_j and θ^i_k, respectively. Then, the degree of simultaneous sampling of topic A^j and topic B^k can be expressed as $t^i \theta^i_j \theta^i_k$, where t^i is the number of words included in document i. Integrating the degree of simultaneous sampling across the entire document set yields

$$D_g(A^j, B^k) = \sum_{i=1}^{N} t^i \theta^i_j \theta^i_k, \quad j = 1, \dots, J, \quad k = 1, \dots, K, \tag{10}$$

which represents the overlap between A^j and B^k. Hence, we obtain a confusion matrix (**Table 1**) with $D_g(A^j, B^k)$ as an element. The total degree, say T, equals the number of words in the total of documents $\sum_{i=1}^{N} t^i$, since $\sum_j \theta^i_j = \sum_k \theta^i_k = 1 \ \forall \ i$.

Table 1 is a frequency distribution. Dividing this by the total frequency yields a two-dimensional probability distribution. The simultaneous probability is given by

$$P(A^j, B^k) = \frac{1}{T} D_g(A^j, B^k). \tag{11}$$

Using the simultaneous probabilities $P(A^j, B^k)$, the NMI between two solutions (discrete random variables) A and B is defined as [5]

	B^1	B^2	$\cdots$	B^K	Sum
A^1	19.1	2.3	$\cdots$	7.4	312.3
A^2	24.2	0.2	$\cdots$	4.9	319.5
$\vdots$	$\vdots$	$\vdots$	$\cdots$	$\vdots$	$\vdots$
A^J	1.2	4.8	$\cdots$	19.0	293.6
Sum	213.3	196.7	$\cdots$	253.2	T

Table 1.
Confusion matrix.

$$\mathrm{NMI}(A, B) = \frac{I(A;B)}{(H(A) + H(B))/2}, \tag{12}$$

where $I(A;B)$ is the mutual information and $H()$ is the entropy. Specifically

$$I(A;B) = \sum_{j=1}^{J} \sum_{k=1}^{K} P(A^j, B^k) \log \frac{P(A^j, B^k)}{P(A^j)P(B^k)}, \tag{13}$$

$$H(A) = \sum_{j=1}^{J} -P(A^j) \log P(A^j), \quad H(B) = \sum_{k=1}^{K} -P(B^k) \log P(B^k). \tag{14}$$

Since a symmetrical relationship $\mathrm{NMI}(A, B) = \mathrm{NMI}(B, A)$ is satisfied from above, NMI can be thought of as a similarity related to information and also as an inner product between solutions.

Let $\{\boldsymbol{\theta}_l | l = 1, \dots, L\}$ be a set of solutions about topic models. Forming the inner product matrix $\boldsymbol{B}(L \times L)$ whose elements are the inner products calculated as NMI between them, then the matrix $\boldsymbol{B}$ enables us to assign coordinate values to the solutions by multidimensional scaling (MDS) [6].

Let $\boldsymbol{y}_l$, $\boldsymbol{Y} = (\boldsymbol{y}_1 \dots \boldsymbol{y}_L)$ be the coordinate value (vector) of l-th solution in Euclidean space, and the matrix expression of them, respectively. $\boldsymbol{B}$ as the inner product matrix can be expressed by

$$\boldsymbol{B} = \boldsymbol{Y}^t \boldsymbol{Y}. \tag{15}$$

Since $\boldsymbol{B}$ is symmetric and positive semidefinite from the definition of NMI, there exists an orthogonal matrix Φ such that

$$\Phi^t \boldsymbol{B} \Phi = \Lambda, \tag{16}$$

where Λ is a diagonal matrix whose elements are eigenvalues of $\boldsymbol{B}$. Hence,

$$\boldsymbol{B} = \Phi \Lambda \Phi^t = \left(\Lambda^{1/2} \Phi^t\right)^t \left(\Lambda^{1/2} \Phi^t\right) = \boldsymbol{Y}^t \boldsymbol{Y}, \tag{17}$$

and we obtain

$$\boldsymbol{Y} = \Lambda^{1/2} \Phi^t. \tag{18}$$

The vectors $\boldsymbol{Y}$ can be used to visualize the solutions in low-dimensional space. The eigenvalue decomposition in (Eq. (17)) is for the origin viewpoint and not the center of gravity, so when applying principal component analysis (PCA), it should again be applied to the vectors $\boldsymbol{Y}$.

4. An analysis method using word distribution

This section introduces the method [7] for representing typical solutions and the topics contained therein, as well as possible topics that can be extracted, by means of the word distribution $\boldsymbol{\phi}$ that are easy for human to understand. Considering all word distributions in diverse solutions, the number of combinations of word distributions is enormous. It is inefficient to treat "quite similar" and "different but similar" relationship equally in order to get a complete picture of word distributions. Therefore, we represent the former relationships by grouping them together under the same representative word distribution and the latter relationships by analysis based on similarity relations among representative word distributions. These are described in detail below.

Information-theoretic clustering [16, 17] based on similarity of probability distributions is used to estimate representative word distributions from word distributions $\{\boldsymbol{\phi}^1, \dots, \boldsymbol{\phi}^i, \dots, \boldsymbol{\phi}^N\}$ included in solutions. The objective function to be minimized in this clustering is expressed as

$$\mathrm{JS_W} = \frac{1}{N}\sum_{f=1}^{K_f}\sum_{\boldsymbol{\phi}^i \in C^f} D_{\mathrm{KL}}(\boldsymbol{\phi}^i \| Q^f) = \frac{1}{N}\sum_{f=1}^{K_f}\sum_{\boldsymbol{\phi}^i \in C^f}\sum_{m=1}^{M} \phi^i_m \log \frac{\phi^i_m}{q^f_m}, \tag{19}$$

where $D_{\mathrm{KL}}()$ denotes Kullback-Leibler divergence, C^f is the estimated cluster of word distributions that are quite similar to each other, and Q^f is the estimated representative word distribution, which is literally representative of the word distributions that could be extracted in topic modeling. The number of clusters K_f should be large enough so that they consist of word distributions that are quite similar to each other.

Representative word distributions with similar relationships are then connected to form *similarity network of representative word distributions*. Whether representative word distributions $\boldsymbol{\phi}^i$ and $\boldsymbol{\phi}^j$ are similar or not is determined by Jensen Shannon (JS) divergence given as

$$D_{\mathrm{JS}}(\boldsymbol{\phi}^i, \boldsymbol{\phi}^j) = D_{\mathrm{JS}}(\boldsymbol{\phi}^j, \boldsymbol{\phi}^i) = \frac{1}{2}\left(D_{\mathrm{KL}}\left(\boldsymbol{\phi}^i \middle\| \frac{\boldsymbol{\phi}^i + \boldsymbol{\phi}^j}{2}\right) + D_{\mathrm{KL}}\left(\boldsymbol{\phi}^j \middle\| \frac{\boldsymbol{\phi}^i + \boldsymbol{\phi}^j}{2}\right)\right). \tag{20}$$

In this network, there are groups of representative word distributions that are similar to each other in areas of high edge density. These groups are extracted by the clustering algorithm based on maximizing modularity [18].

Since word distributions belong to one of the groups via the representative word distribution, solutions can be typified by the frequency distribution of the group to which the included word distribution belongs. As a result, several typical solutions can be found by the analysis using word distributions.

5. Experiments

We used three text data sets: NYtimes [19], 20News [20], Nips [19]. Stop-words included in 20News and documents with fewer than 40 words were removed. The characteristics of data sets actually used are shown in **Table 2**.

Each data set was separated into a training set for 90% of its documents and a test set for 10%. The test set were used to evaluate parameter estimation. The parameters $\boldsymbol{\theta}, \boldsymbol{\phi}$, which are the solutions estimated for the training set, were analyzed. The methods used are (1) MAP estimation (MapRnd), (2) MAP estimation using information-theoretic clustering results for initial value setting (MapCL), (3) collapsed Gibbs sampling (CGS), and (4) collapsed variational Bayesian estimation

Name	Number of documents (N)	Number of word types (M)
NYtimes	296,829	101,631
20News	14,111	60,149
Nips	1491	12,375

Table 2.
Characteristics of data sets used in experiments.

(CVB0). Of these, random initial values were given to the parameters in all cases except (2), where a small value was added to each element of the word distribution of the cluster obtained by weighted information-theoretic clustering to make the initial word distributions ϕ and a uniform distribution was set as the initial topic distribution θ [16].

The number of topics was set to $K = 10$, and the hyperparameters were set to $\alpha = 1.01$ and $\beta = 1.1$ for MAP estimation and $\alpha = 0.01$ and $\beta = 0.1$ for the other methods, adjusting for differences in the update equations [15]. For each method, parameter estimation was performed for 200 different random number series.

5.1 Experimental results

We evaluated a total of 800 solutions $\{\boldsymbol{\theta}, \boldsymbol{\phi}\}$, estimated and acquired through the four methods, by perplexity calculated using the test set (**Table 3**). With the acquired $\boldsymbol{\phi}$ as known, the topic distribution θ_{test} was estimated using half of the words in each document in the test set, and the perplexity was calculated from the word and topic distributions $\boldsymbol{\phi}, \boldsymbol{\theta}_{test}$ using the other half of the words. Since the objective is to evaluate the goodness of solution, the estimation of the topic distribution in the test set is the same for all methods, and MAP estimation was used in this case. **Table 3** shows that MapCL, which uses clustering results as initial values, is superior to MapRnd, and that Cvb0 performs better than the other methods except MapCL. The fact that Cvb0 shows better results is consistent with the results in [15].

We assigned coordinate values to the solutions by the analysis using topic distributions and applied principal component analysis PCA to visualize them in **Figure 3**.

In **Figure 3**, we see that MapCL is biased toward the range of large (20News) and small (Nips) values of the first principal component. This indicates that the solutions are method-dependent. These diagrams are useful to get an overall picture of the distribution of solutions, and by choosing solutions far from each other (e.g., top left, top right, center, bottom left, bottom right), a non-redundant solution set is obtained. However, how they differ is difficult for humans to understand. Therefore, it is important to analyze using word distributions that are easy for humans to understand.

Name	MapRnd	MapCL	Cgs	Cvb0
NYtimes	5983.8	5959.3	5991.7	5973.7
20News	5693.3	5662.4	5772.7	5688.9
Nips	1993.3	1989.3	2006.4	1982.3

Table 3.
Perplexities archived in the four methods.

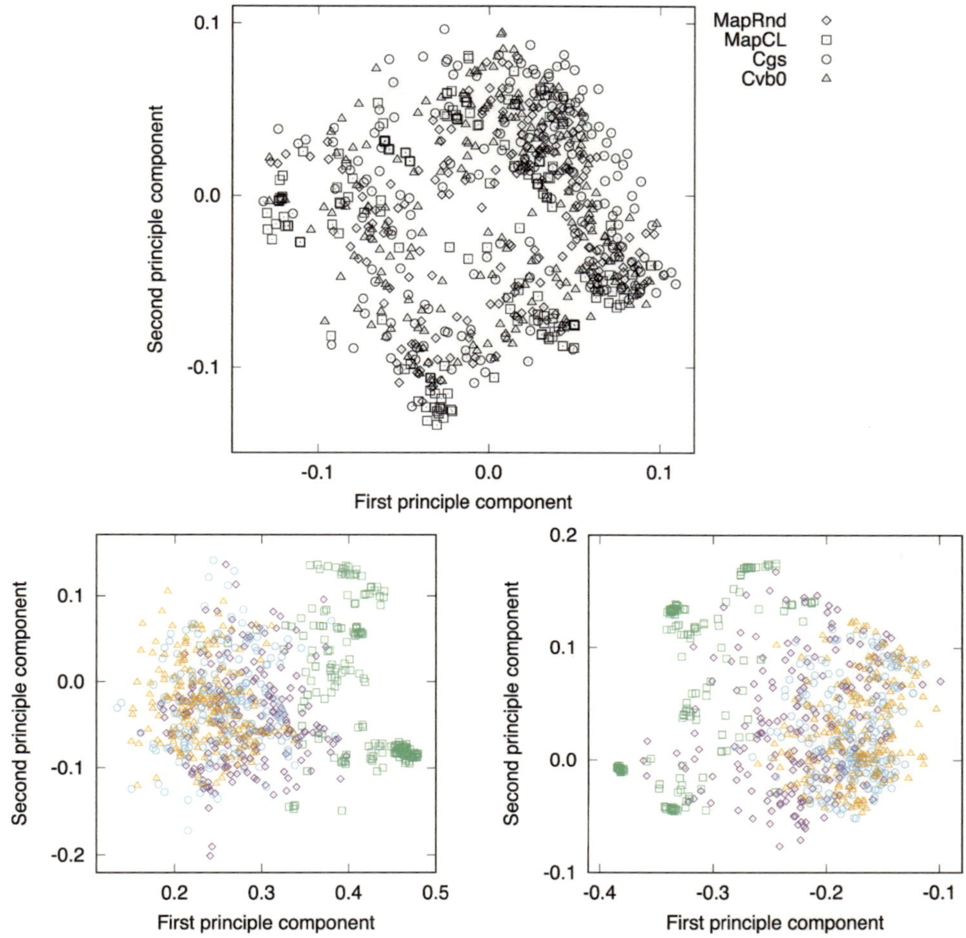

Figure 3.
Visualizations based on principal component analysis of solutions for NYtimes (top), 20News (bottom left), and Nips (bottom right).

For the analysis using word distributions, 100 representative word distributions were first obtained by information-theoretic clustering (see Eq. (19)) from a total of 8000 word distributions, 10 in each solution. The frequency distributions of JS divergence between word distributions and JS divergence between representative word distributions are shown in **Figure 4 (left)**. Since the two frequency distributions are well matched, and the representative word distributions preserve the relationships among the word distributions, $K_f = 100$ would be sufficient for the number of clusters.

Figure 4 (right) shows the frequency distribution of the JS divergence between word distributions inside each solution and between representative word distributions corresponding to the word distributions. The word distributions in the solution of topic models are estimated to be different from each other in the sense of optimizing the objective function (Eq. (2)). Therefore, the JS divergence between word distributions inside the solution has a larger value. Considering this figure, we determined that word distributions are similar to each other if the JS divergence between them is equal to 0.1 or less.

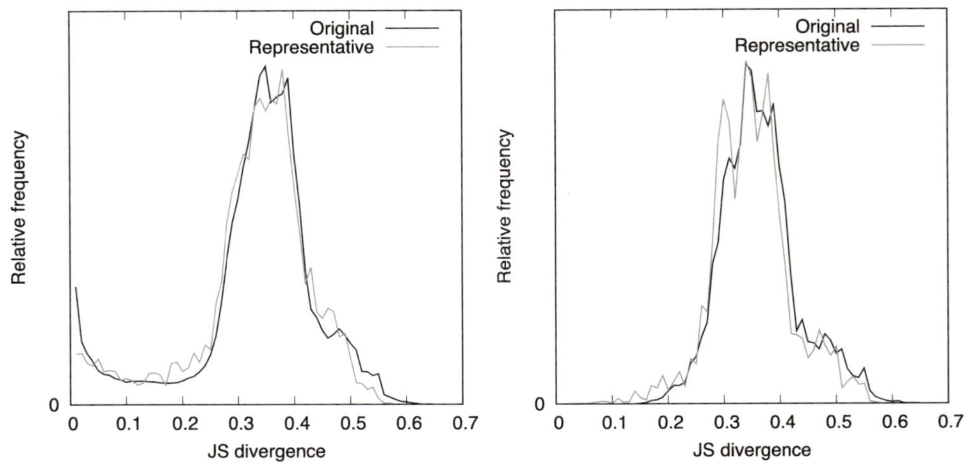

Figure 4.
Frequency distribution of JS divergence between word distributions for all solutions (left) and inside solutions (right).

We then connected representative word distributions with similar relationships for the NYtimes data set and represented them as a similarity network of representative word distributions in **Figure 5**. From this network, the clustering algorithm based on maximizing modularity [18] was used to extract groups (g1 to g15) with representative word distributions that were in regions of high edge density. There were 10 large groups with four or more vertices, the same as the number of topics K. In **Figure 5**, the vertices are colored to distinguish the groups and only the network information

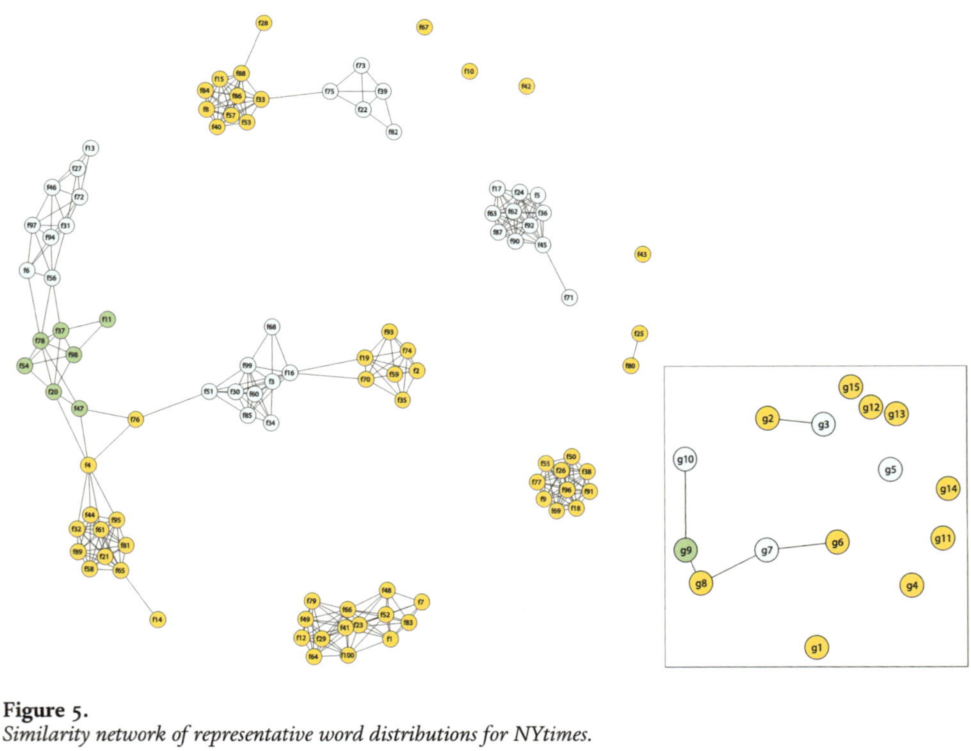

Figure 5.
Similarity network of representative word distributions for NYtimes.

representing the adjacencies is meaningful, not the positions (coordinate values) of the vertices.

Table 4 shows the high-frequency words in the representative word distributions for each large group (g1 to g10) in NYtimes. For the adjacent and ambiguous groups (g6 to g10) in **Figure 5**, two representative word distributions were chosen to represent the variation within the group, and the characteristic words representing the differences are shown in bold. The words "school student" appear in g6, g7, and g8, suggesting that there are a variety of topics related to these words. We see that the word "drug" is listed with "doctor" in g6, but with "case" in g7.

The representative word distributions within these groups are somewhat different, and it is not easy to select the appropriate one. Therefore, the proposed method of representing relationships in a human-understandable form may be useful for users. If we were to name the groups according to the high-frequency words, they would be, in order, **sports (g1), markets (g2), IT (g3), presidential election (g4), international conflicts (g5), health care (g6), school (g7), entertainment (g8), housing (g9),** and **food (g10)**.

We typified solutions by the frequency distribution of the group to which the word distribution in the solution belongs. We call the types of frequency distributions *patterns*, and the top five most frequently occurring patterns are listed in **Table 5**. As the tables show, these patterns consist of combinations of the large groups (g1 to g10). For patterns 1 and 2, we selected solutions that are typical in the sense that we often find combinations of representative word distributions associated with the word distributions in the solution, and listed in **Table 6** the high-frequency words in the word distributions belonging to these solutions. In the tables, the names of the groups to which the word distributions belong are indicated.

gn	fn	High-frequency words
g1	f41	team game season player play games point run coach win won right hit left
g2	f40	company percent million companies market stock business billion money
g3	f22	com web computer site information www mail online .internet internet
g4	f55	.bush president campaign .george_bush .al_gore election political vote
g5	f36	official government .united_states .u_s military war attack palestinian leader
g6	f2	**drug patient doctor** cell research problem scientist percent health study
	f70	school student drug patient percent program doctor women study problem
g7	f16	**school case student drug** law court official patient children lawyer found
	f51	**school student family** children case home told police death law lawyer official
g8	f76	school student children family women home friend father mother parent
	f61	**show book film** movie look music friend women play family character love
g9	f20	home family room building friend night house children town .new_york
	f37	**car building** water room home air hour area miles town house feet place
g10	f6	water food room cup minutes small building hour restaurant add large home
	f72	**food cup** minutes add water oil restaurant wine tablespoon fat sugar chicken

Table 4.
High-frequency words in the representative word distribution for each group in NYtimes.

As **Table 6 (top)** shows, the pattern 1 has no topics on IT and instead has two topics on sports. The pattern 3 has no topics related to housing in g9 (see **Table 5**). If we were to have the right set of topics for a smart city, we should choose a solution from the pattern 2 (see **Table 6 (bottom)**), which includes the all groups, rather than focusing on sports.

	g1	g2	g3	g4	g5	g6	g7	g8	g9	g10	Number
Pattern 1	2	1	0	1	1	1	1	1	1	1	125
Pattern 2	1	1	1	1	1	1	1	1	1	1	81
Pattern 3	2	1	1	1	1	1	1	1	0	1	67
Pattern 4	2	1	1	1	1	0	1	1	1	1	66
Pattern 5	1	1	0	1	1	1	1	2	1	1	46

Table 5.
High-frequently occurring patterns. "Number" indicates the number of solutions belonging to the pattern.

Group name	High-frequency words
Sports	season team game run games hit player inning play baseball right
Sports	team game season player play point games coach win won yard
Markets	company percent million companies market business stock billion
Presidential election	.bush president campaign .george_bush .al_gore election political
International conflicts	official government .united_states attack .u_s military war leader
Health care	drug patient doctor percent problem cell research study health
School	school student case law court lawyer children police official family
Entertainment	show book film movie music look play women friend character
Housing	car building home room water hour house area air town miles
Food	cup food minutes add oil tablespoon wine sugar pepper water
Group name	**High-frequency words**
Sports	team game season player play games point run coach win football
Markets	percent company million companies market stock business billion
IT	com web computer information site www mail online .internet
Presidential election	.bush campaign president .george_bush .al_gore election political
International conflicts	official government .united_states attack .u_s military war leader
Health care	drug patient doctor health research study scientist cell problem
School	school student law case court lawyer official children police family
Entertainment	show film book movie look women play music friend character
Housing	car building home room water hour house town miles area air
Food	cup food minutes add oil wine tablespoon sugar pepper

Table 6.
High-frequency words in typical solutions for pattern 1 (top) and pattern 2 (bottom).

Figure 6 shows similarity networks of representative word distributions for the 20News and Nips data sets. As these figures show, for both data sets, the number of large groups was 10, the same as the number of topics.

The 20News and Nips solutions were typified based on the frequency distribution of the groups, and the top five most frequently occurring patterns are shown in **Table 7**. As the tables show, these patterns also consist of combinations of the large

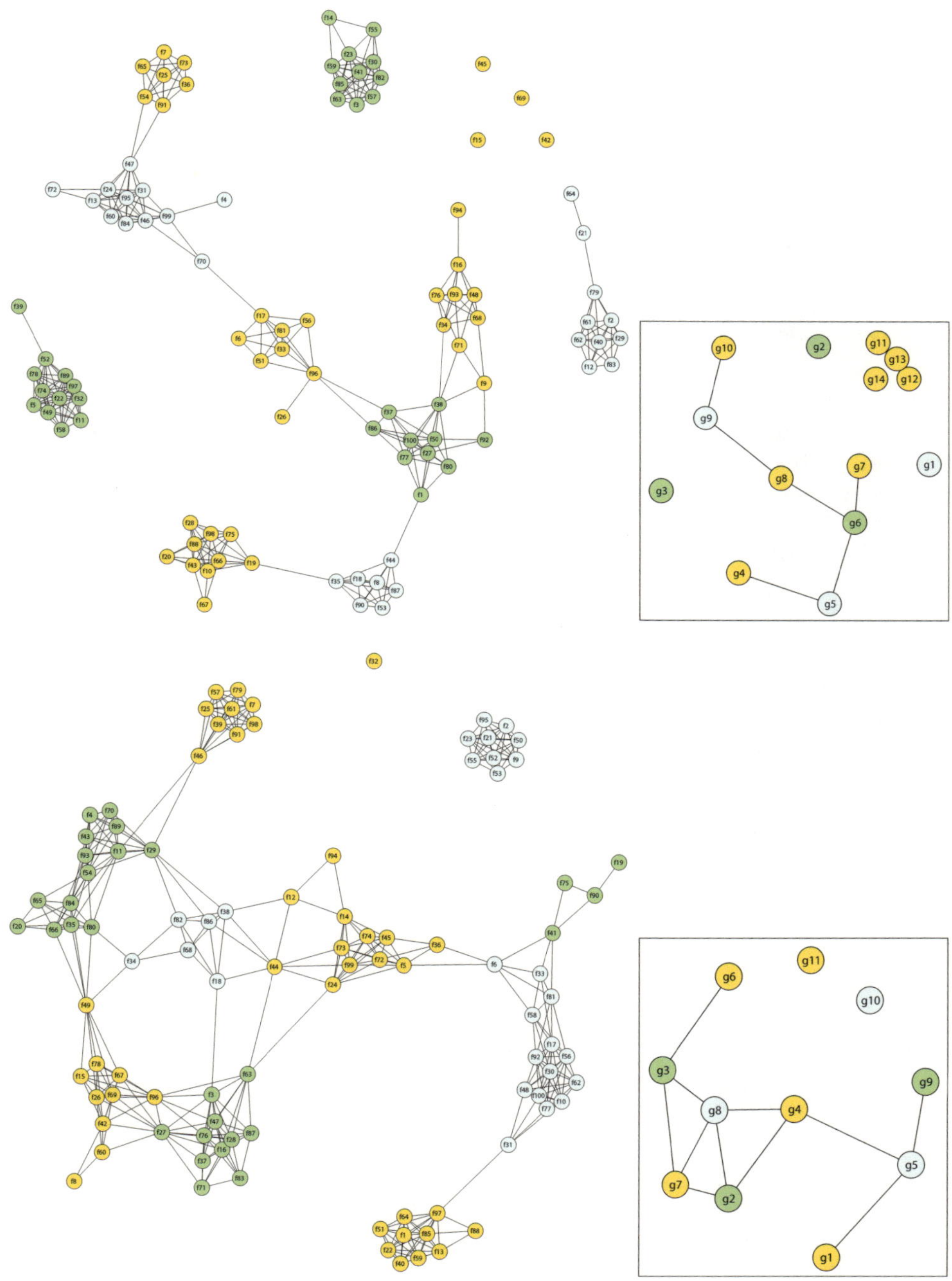

Figure 6.
Similarity networks of representative word distributions for 20News (top) and Nips (bottom).

	g1	g2	g3	g4	g5	g6	g7	g8	g9	g10	Number
Pattern 1	1	1	1	1	1	1	1	1	1	1	215
Pattern 2	2	1	1	1	1	1	1	0	1	1	44
Pattern 3	1	2	1	1	1	1	1	0	1	1	39
Pattern 4	1	2	1	1	0	1	1	1	1	1	29
Pattern 5	1	1	1	1	1	1	2	0	1	1	27
	g1	**g2**	**g3**	**g4**	**g5**	**g6**	**g7**	**g8**	**g9**	**g10**	**Number**
Pattern 1	1	1	2	1	2	1	1	0	0	1	119
Pattern 2	1	1	2	1	1	1	1	0	1	1	81
Pattern 3	1	1	1	1	2	1	1	1	0	1	67
Pattern 4	1	1	1	1	1	1	1	1	1	1	64
Pattern 5	1	1	2	1	1	1	1	1	0	1	56

Table 7.
High-frequently occurring patterns in 20News (top) and Nips (bottom).

groups (g1 to g10). Since large groups play a role in many solutions, a solution with one word distribution for all large groups would be the solution of interest. However, such a solution is not necessarily the most common solution, nor is it necessarily the optimal solution. For reference, **Table 8** shows examples of such solutions. The 20News example belongs to the most common pattern (pattern 1) and could be a candidate for a good solution. In the 20News data set, documents are labeled with the newsgroup to which they belong [20]. Using the label information, we can find the high-frequency words of the documents belonging to each newsgroup (see **Table 9**). Note that topic modeling does not use label information. Comparing the topics in **Table 8 (top)** with those in **Table 9**, many of them are associated. For example, g1 is associated with n11, g3 with n8–9, g4 with n15, g5 with n14, and g9 with n2. This association with the actual newsgroups would supports that the solution in **Table 8 (top)** is appropriate. The Nips example belongs to the fourth most common pattern, and a solution to the other patterns would be appropriate.

Group	High-frequency words
g1	game team writes year games article hockey play good players season ca time win
g2	god people jesus writes article christian bible church christ time good life christians
g3	writes car article good apr bike dod ve time ca cars back engine ll make front thing
g4	space nasa writes earth article gov launch system time orbit science shuttle moon
g5	writes article people health medical disease time cramer patients cancer study doctor
g6	people writes article government gun president make fbi time mr state law fire guns
g7	people israel armenian jews turkish writes armenians article war israeli jewish arab
g8	key encryption chip writes government information clipper keys system article
g9	file image window program ftp windows files graphics version server jpeg display
g10	dos windows drive writes card scsi system article mb pc problem mac disk bit work

Group	High-frequency words
g1	circuit signal analog chip system output current neural input neuron voltage filter
g2	speech word recognition system model context network hmm training sequence set
g3	function learning algorithm point vector result error bound case equation set
g4	image object images recognition feature map features representation set vector point
g5	neuron cell model input pattern network activity synaptic visual stimulus firing
g6	model data distribution parameter gaussian algorithm function method mean
g7	network unit input weight neural output learning training hidden layer error net
g8	set training data error algorithm classifier performance learning classification test
g9	model control system motion direction position movement motor eye learning field
g10	learning action function algorithm problem reinforcement policy system optimal

Table 8.
High-frequency words in solutions having one word distribution for all large groups for 20News (top) and Nips (bottom).

Newsgroup	High-frequency words
(n1) alt.atheism	god writes people article atheism religion time evidence jesus
(n2) comp.graphics	image graphics jpeg file bit images software data files ftp format
(n3) comp.os.ms-windows.misc	windows file dos writes article files ms os problem win program
(n4) comp.sys.ibm.pc.hardware	drive scsi card mb ide system controller bus pc writes disk dos
(n5) comp.sys.mac.hardware	mac apple writes drive system problem article mb monitor mhz
(n6) comp.windows.x	window file server windows program dos motif sun display
(n7) misc.forsale	sale shipping offer mail price drive condition dos st email
(n8) rec.autos	car writes article cars good engine apr ve people time ford speed
(n9) rec.motorcycles	writes bike article dod ca apr ve ride good time bmw back riding
(n10) rec.sport.baseball	writes year article game team baseball good games time hit players
(n11) rec.sport.hockey	game team hockey writes play ca games article season year nhl
(n12) sci.crypt	key encryption government chip writes clipper people article keys
(n13) sci.electronics	writes article power good ve work ground time circuit ca make
(n14) sci.med	writes article people medical health disease time cancer patients
(n15) sci.space	space writes nasa article earth launch orbit shuttle time system
(n16) soc.religion.christian	god people jesus church christ writes christian christians bible
(n17) talk.politics.guns	gun people writes article guns fbi government fire time weapons
(n18) talk.politics.mideast	people israel armenian writes turkish jews article armenians israeli
(n19) talk.politics.misc	people writes article president government mr stephanopoulos
(n20) talk.religion.misc	god writes people jesus article bible christian good christ life time

Table 9.
High-frequency words in the 20News data set.

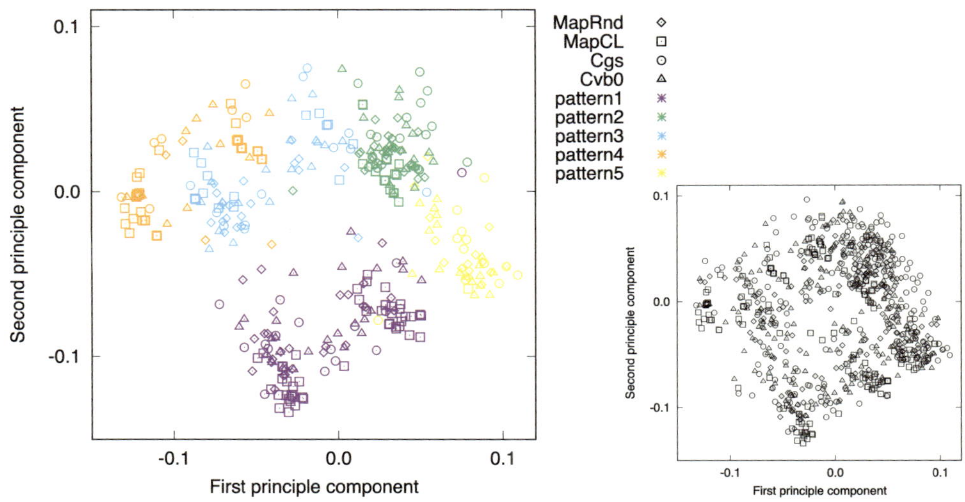

Figure 7.
Visualization based on principal component analysis of solutions in NYtimes. Patterns 1–5 (left), for all (right).

In order to select an appropriate solution, it is crucial to determine that the solution is consistent with the objective. Selecting a pattern that matches the objective would be the first step in finding a solution. After that, the solution can be obtained by refining the word distributions that the solution should have by selecting representative word distributions for each group, while confirming the objective.

Since determining the degree of consistency is a relative issue, it is essential to know the overall picture of the solutions and possible word distributions in them. **Figure 7** shows a visualization of the solutions using the patterns assigned by the analysis based on word distribution and the coordinate values assigned by the analysis based on topic distribution. **Figure 7 (left)** is more useful than **Figure 7 (right)** (equivalent to **Figure 3 (top)**) in finding a solution, as it provides additional information on the patterns needed in the first step.

The solutions for patterns 1–5 in **Figure 7 (left)** are grouped together in each pattern. It means that the solutions with the same pattern are also similar to each other in topic distribution. It is interesting to confirm that word and topic distributions are related.

As the experimental results show, analysis using word distribution will play a major role in the search for a solution. This is because the results can be presented in a way that is understandable to humans and can be compared to the objective. Analysis using topic distribution will play the role of a "map" that provides another point of view when a decision is not clear. A map that provides a view of all the solutions should be useful.

6. Conclusion

It has been reported that information-theoretic clustering outperforms spherical clustering when targeting text data [17]. Topic modeling is an extension of information-theoretic clustering (see Appendix A), which is why we apply this technique to document analysis. The solutions obtained through modeling are diverse.

In past studies, however, diversity has not been adequately considered. This chapter introduced methods for analyzing diverse solutions and obtaining an overall picture of the solutions. Also, we showed effectiveness of the methods through experiments.

In this study, we found that there are many solutions that are different from each other in topic models. It is difficult to obtain an appropriate solution by chance. Furthermore, problems in the world, not to mention smart cities, are complex and change rapidly, so there is a high risk of missing important topics. The proposed analysis methods should be useful in the search for solutions. There are various extensions of topic models, such as dynamic topic models [21], but even there, a diversity of solutions may exist. The approach presented in this chapter may also be used for analyses using such models.

Appendix A

We present the objective function of weighted information-theoretic clustering (ITC) [16, 17] and show that it is a special case of the objective function of topic models.

Assume that there are *cluster* $C^k (k = 1, \dots, K)$ and observed value vector $\boldsymbol{x}^i (i = 1, \dots, N)$, and that there are M types of words in the total of observed value vectors. Also, assume that each *cluster k* has a word distribution $\boldsymbol{\phi}^k = \{\phi_1^k, \dots, \phi_m^k, \dots, \phi_M^k\}$ and vector $\boldsymbol{x}^i$ belong to one of the clusters. Clusters in clustering is regarded as the same concept as topics. The objective function of ITC $\mathrm{JS_W}$ and that of weighted ITC $\mathrm{JS'_W}$ are given by

$$\mathrm{JS_W} \propto \sum_{k=1}^{K} \sum_{\mathbf{x}^i \in C^k} D_{\mathrm{KL}}\left(\boldsymbol{p}^i \| \boldsymbol{\phi}^k\right) = \sum_{k=1}^{K} \sum_{\boldsymbol{x}^i \in C^k} \sum_{m=1}^{M} p_m^i \log \frac{p_m^i}{\phi_m^k}, \tag{21}$$

$$\mathrm{JS'_W} \propto \sum_{k=1}^{K} \sum_{\boldsymbol{x}^i \in C^k} t^i D_{\mathrm{KL}}\left(\boldsymbol{p}^i \| \boldsymbol{\phi}^k\right) = \sum_{k=1}^{K} \sum_{\boldsymbol{x}^i \in C^k} \sum_{m=1}^{M} x_m^i \log \frac{p_m^i}{\phi_m^k}, \tag{22}$$

where $\boldsymbol{p}^i = \boldsymbol{x}^i / \|\boldsymbol{x}^i\|_1$ and $t^i = \|\boldsymbol{x}^i\|_1$ denote the word distribution and l^1-norm of the i-th vector $\boldsymbol{x}^i$, respectively. Comparing both objective functions, $\mathrm{JS'_W}$ is *weighted* by t^i equal to the number of words in the i-th data (document). This is because it treats the occurrence of words equally, as in topic models, whereas normal clustering treats each document equally.

The interior of Eq. (22) can be transformed as

$$x_m^i \log \frac{p_m^i}{\phi_m^k} = \left(-x_m^i \log \phi_m^k\right) - \left(-x_m^i \log p_m^i\right), \tag{23}$$

where the second term is independent of the clustering result. Thus, the function to be minimized can be expressed as

$$\sum_{k=1}^{K} \sum_{\boldsymbol{x}^i \in C^k} \sum_{m=1}^{M} \left(-x_m^i \log \phi_m^k\right). \tag{24}$$

Meanwhile, the function to be maximized in topic models is

$$\sum_{i=1}^{N}\sum_{m=1}^{M} x_m^i \log\left(\sum_{k=1}^{K}\theta_k^i\phi_m^k\right), \tag{25}$$

which is the second term in Eq. (2). Applying the hard clustering constraint:

$$\theta_k^i = \begin{cases} 1 & x^i \in C^k \\ 0 & \text{otherwise} \end{cases}, \tag{26}$$

we obtain

$$\sum_{k=1}^{K}\sum_{x^i \in C^k}\sum_{m=1}^{M} x_m^i \log\phi_m^k. \tag{27}$$

Since minimization of Eq. (24) is equivalent to maximization of Eq. (27), topic modeling includes weighted ITC as a special case and is an extension of it.

For weighted ITC, the learning algorithm needs to be changed from ITC [17]. Basically, it should treat documents differently based on the number of words they contain. There are two ways to achieve this: one is to select documents with a probability proportional to the number of words they contain, and the other is to increase the learning rate of competitive learning in proportion to the number of words in the documents selected at learning time. In this experiment, we employed the latter [16].

Acknowledgements

This work was supported by JSPS KAKENHI Grant Number 18 K11442.

Author details

Toshio Uchiyama* and Tsukasa Hokimoto
Hokkaido Information University, Ebetsu-shi, Hokkaido, Japan

*Address all correspondence to: uchiyama.toshio@do-johodai.ac.jp

References

[1] Hofmann T. Unsupervised learning by probabilistic latent semantic analysis. Machine Learning. 2001;**42**(1–2):117-196

[2] Blei DM, Ng AY, Jordan MI. Latent dirichlet allocation. Journal of Machine Learning Research. 2003;**3**:993-1022

[3] Uchiyama T. A method for analyzing solution diversity in topic models. In: Proceedings of 5th International Conference on Business and Industrial Research (ICBIR). New York: IEEE; 2018. pp. 29-34. DOI: 10.1109/ICBIR.2018.8391161

[4] Uchiyama T, Hokimoto T. Analysis and visualization of solution diversity about topic model. IEICE Transactions D. 2019;**J102-D**(10): 698-707. DOI: 10.14923/transinfj.2019JDP7017

[5] Manning CD, Raghavan P, Schütze H. Introduction to Information Retrieval. Cambridge, England: Cambridge University Press; 2008

[6] Torgerson WS. Multidimensional scaling: I theory and method. Psychometrika. 1952;**17**(4):401-419

[7] Uchiyama T, Hokimoto T. A word distribution based analysis of the diverse solution at topic model. IEICE Transactions D. 2022;**J105-D**(5):405-415. DOI: 10.14923/transinfj.2021JDP7053

[8] Gondek D, Hofmann T. Non-redundant data clustering. Knowledge and Information Systems. 2007;**12**(1):1-24

[9] Niu D, Dy JG, Jordan MI. Multiple non-redundant spectral clustering views. In: Proceedings of the 27th International Conference on Machine Learning (ICML-10). Madison, Wisconsin, United States: Omnipress; 2010. pp. 831-838

[10] Caruana R, Elhawary M, Nguyen N, Smith C. Meta clustering. In: Proceedings of Sixth International Conference on Data Mining. New York: IEEE; 2006. pp. 107-118

[11] Iwata T, Yamada T, Ueda N. Probabilistic latent semantic visualization: Topic model for visualizing documents. In: Proceedings of the 14th ACM SIGKDD International Conference on Knowledge Discovery and Data Mining. New York: ACM; 2008. pp. 363-371

[12] Griffiths TL, Steyvers M. Finding scientific topics. In: Proceedings of the National Academy of Sciences. Vol. 101. Washington, D.C.: National Academy of Sciences; 2004. pp. 5228-5235

[13] Teh YW, Newman D, Welling M. A collapsed variational Bayesian inference algorithm for latent dirichlet allocation. In: Proceedings of Advances in Neural Information Processing Systems. Cambridge, MA, United States: MIT Press; 2007. pp. 1353-1360

[14] Chien JT, Wu MS. Adaptive Bayesian latent semantic analysis. IEEE Transactions on Audio, Speech, and Language Processing. 2008;**16**(1): 198-207

[15] Asuncion A, Welling M, Smyth P, Teh YW, Asuncion A, Max W, et al. On smoothing and inference for topic models. In: Proceedings of the Twenty-Fifth Conference on Uncertainty in Artificial Intelligence. Arlington, Virginia, United States: AUAI Press; 2009. pp. 27-34

[16] Uchiyama T. Improvement of probablistic latent semantic analysis using information-theoretic clustering.

IEICE Transactions D. 2017;**J100-D**(3): 419-426. DOI: 10.14923/transinfj.2016JDP7085

[17] Uchiyama T. Information theoretic clustering and algorithms. In: Hokimoto T, editor. Advances in Statistical Methodologies and Their Application to Real Problems. London, UK: IntechOpen; 2017. pp. 93-119. DOI: 10.5772/66588

[18] Newman MEJ, Girvan M. Finding and evaluating community structure in networks. Physical Review E. 2004; **69**(2):026113

[19] Dua D, Graff C. UCI Machine Learning Repository. Available from: http://archive.ics.uci.edu/ml/ [Accessed: 16 May 2022]

[20] 20news-bydate data at Home page for 20 Newsgroups data set. Available from: http://qwone.com/jason/20Newsgroups/20news-bydate-matlab.tgz [Accessed: 16 May 2022]

[21] Blei DM, Lafferty JD: Dynamic topic models. In: Proceedings of the 23rd International Conference on Machine Learning. New York: Association for Computing Machinery; 2006. pp. 113–120

Chapter 4

Information and Communication Technologies for New Generation of Sustainable Smart Cities

Kamal Shahid, Muhammad Hassan, Ali Husnain and Sadaf Mukhtar

Abstract

The huge growth of population in cities all over the world has forced countries to regulate and manage resources in these places. Therefore, urban waste management, fossil fuel conservation, affordable and resourceful healthcare systems, effective traffic management, government transparency, and other concerns plague the world's main cities. These issues have prompted the creation of Sustainable Smart Cities, which are innovative, technology-based, and environmentally friendly urban areas. The sustainable smart cities deploy technologies specifically Information and Communication Technologies (ICT) to keep an eye on the community and develop long-term, cost-effective solutions. Thus, for the effective implementation of sustainable smart cities, a stable, secure, inter-operable, and reliable telecommunication network is necessary to enable applications and services in metropolitan areas. Recent advancements in the areas of 5G, 6G, Block chain technology, Internet of Things (IoT), and Artificial Intelligence (AI) are anticipated for working and assisting the creation of sustainable smart cities. This chapter provides an introduction of the elements of sustainable smart cities, as well as an overview of how cities throughout the world have adopted them and projected trends for the next generation of sustainable smart cities.

Keywords: sustainable, smart cities, IoT, blockchain, 5G, 6G

1. Introduction

A city that is smart, sustainable, and innovative is a city that is smart, sustainable, and inventive that employs Information and Communication Technologies (ICT) as well as other ways to raise the standard of living, the efficiency with which urban activities and services are carried out, and the competitiveness of the city while also meeting the financial, societal, environment, and historical needs of the growing generations. Many individuals are migrating from rural to urban regions in search of better employment and health. Smart cities rely heavily on information and communication technology. It increases residents' well-being by providing better services. Smart cities are efficient while also controlling complexity. The economy increases at a steady rate in tandem with the increasing rise of cities. People are increasingly investing in this field.

In every way, ICT is critical. Cities' challenges can be solved through information and communication technologies. They also make certain that they are both ecologically friendly and cost-effective. Water management, electricity, solid waste, public transportation, traffic, and congestion are all areas where ICT can help. ICT is a crucial platform of a smart sustainable city is to establish an intelligent and cost-effective metropolitan setting without compromising the luxury, ease, or standard of living of its residents. ICT is a crucial platform for connecting a wide range of everyday resources to public infrastructure, such as resources, transportation, and water [1].

ICT is a critical component that allows different domains to communicate and facilitates the planning and handling of huge amounts of information, resulting in smartly oriented urban systems and applications, civic engagement, and new services and applications in various aspects of urban life, such as transportation [2]. Given the crucial role of ICT in cities in the coming years, it is critical to create a robust and trustworthy ICT infrastructure that will allow the city to respond more aggressively to future crises while also boosting the quality of ICT, and therefore the people' standard of living [3].

Fifth-generation wireless technology (5G) provides greater system capacity, higher data speeds, much lower latency, higher reliability, and higher communication and excellent information in smart cities. Smart city systems use 5G technology to improve sustainability. The 5G network's strength is tested in all environmental, social, and economic aspects, as well as tiny dimensions like energy efficiency, energy consumption, environmental impact, pollution, cost, health, safety, and security, among others [4].

5G is a vital part of the city's progress, enabling the much-needed infrastructure in smart cities to reach a promising but critical stage. The Smart City concept is feasible, and it is now taking shape in a number of European and international cities. Communities must now support local cellular installations to allow 5th generation communication infrastructure in order for Smart Cities to reach their full potential and reap the full advantages [5].

In addition to the high density of well-informed communication in smart cities, huge system capacity with enormous data speeds, incredibly low latency, and great dependability are all possible with 5G cellular technology. Popular systems including better mobile wavelength services, low-latency reliability, and high-density machine communication are expected to be revolutionized by future networks. This emphasizes the need of researching the long-term sustainability of 5G networks in smart cities in order to both energy efficient and environmentally benign [4].

The readers can find plenty of articles, blogs, and books on the importance of the internet and mobile networks to communities and the global economy. The next generation of these networks, which includes 6G and the Internet of Things (IoT), was recently suggested with the goal of providing city users with seamless communication skills. According to industry projections, the market for smart IoT devices would exceed 50 billion dollars by 2020. Smart applications are likely to spearhead new breakthroughs in cities centered on 6G/IoT as smart IoT devices become more prevalent. By integrating 6G/IoT-based solutions into the ecosystem to innovate, depending on the vision of network infrastructure needed to gather in-depth community information in emerging intelligent communities as well as cities, smart apps play an obvious role in progressing smart cities [6].

In smart cities, IoT technology poses a variety of difficulties, including increasing energy consumption and hazardous and E-waste contamination. Smart city apps must be eco-friendly, which is why they must transition to green IoT. Smart cities become more environmentally safe because of green IoT. As a result, environmental conservation, and cost-cutting measures must all be addressed [7].

IoT is linked to large-scale data analysis, which is reportedly making its way into more metropolitan areas in order to increase energy efficiency and mitigate environmental consequences. This is mostly connected to the optimal use of environmental assets, smart infrastructure and resource management, and enhanced environmental support service delivery. As a result, IoT and big data-related applications can help to construct and improve a sustainable environmental design process [8].

Smart cities may be created in six categories using IoT technology: smart people, economics, transportation, environment, governance, and intelligent living With IoT technology, smart cities may link items, people, and information via computer networks. Sensory issues, including reliability, connectivity, and data storage must be addressed in order to efficiently employ IoT technology on daily basis. Data receivers may have an impact on data gathering samples, numerical factors, and infrastructure results. Thousands of network nodes, such as operating systems, operate together in the Internet of Things, translating the natural world into a compressed form of data. IoT technology isolates data using a tiny cloud computing system that connects gadgets and everyday items via an internet connection extension. Several scientists have attempted to explain the various types of IoT [9].

2. Importance of smart cities

Smart city is a part of your smartness, healthy, and good life. The basic goal of a smart city is to optimize city functions and boost the economy. Using smart technologies and data analysis, we can also improve people quality of life. ICT innovation has always been important to the creation of new cities, particularly smart cities. Many cities have contested the development of ICT, using terms like as intelligent, digital, virtual, and ubiquitous since smart cities were launched [10]. As a result, many smart city studies have emphasized the use of contemporary technologies to improve municipal operations. When looking at a city, there are various dimensions to consider. These dimensions are given in **Figure 1**.

- Infrastructure of smart city which includes Green Energy, Public transport, Emergency services, Technology creativity.

- Environment of the smart city which includes landscape, buildings, parks, lakes, rivers

- Solutions of smart city which include e-traffic, e-government, e-learning,

- Collaboration System of the smart city which includes data collaboration, Open Data, synergy, innovation.

- Living of the smart city which includes Education, work, recovery, playing, Human Resources

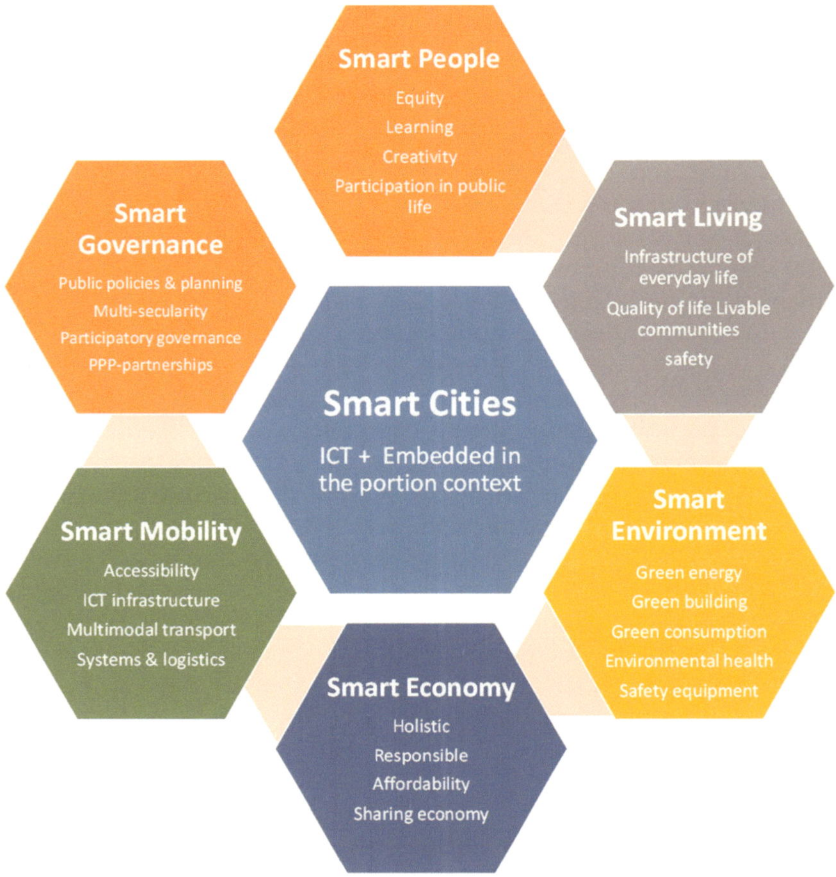

Figure 1.
Dimension of smart city.

2.1 Smart city – infrastructure

A smart city is defined by its ability to bring together people, ideas, resources, knowledge, and technologies to create an efficient, sustainable and strong infrastructure that provides quality services while increasing inhabitants' quality of life. There are many infrastructure plans that are managed and operated by the city. With these infrastructure programs, the city provides services to its residents. The infrastructure system varies from city to city. All major topics connected to a smart city, including as smart travel, smart economics, smart living, smart people smart environment, and smart infrastructure, are built on top of smart government.

Smart mobility includes access to secure travel systems, modern resources, and green infrastructure Means local access, and access to safe, sustainable, and modern transportation systems. Furthermore, smart mobility entails giving individuals access to new technology in order to make the urbanization process easier. Additionally, the current transportation infrastructure should provide access to city travel information via public transportation. The use of ICT to revitalize transportation operations in order to provide accessible mobility is known as smart mobility. As a result, cities should use ICT to increase mobility and develop a digitized and connected transportation network [11].

Smart energy also has an important role in infrastructure. Smart energy management systems using renewable energy sources, sensors, digital controls, advanced meters, automated analysis tools, monitoring, and optimization distribution and use. Such systems improve grid performance and usability of the requirements of the many contributors (producers, suppliers, and consumers). Renewable energy generation, automated demand feedback, micro grids, intelligent grid technology, energy conservation, power plants, and new needs such as electric automobiles and smart electrical goods are all examples of smart energy infrastructure advances. Such innovative approaches enable community-based energy monitoring programs and increase energy efficiency properties by expanding the network of smart power devices across the city and providing a full picture of energy consumption trends.

Smart grids are an important component of smart infrastructure. A smart grid might can be described as a system of supply of electricity from generation to place use combined with ICT to improve grid performance, customer service, and environmental benefits. Smart grids are employed in both wealthy and poor countries across the world. For example, The smart grid used in Japan's Kashiwa-no-ha smart city project is based on a universal energy management system that integrates home power management systems, real-time monitoring of power supply and demand, and self-support energy management with the appropriate amount of energy generated and saved.

The idea of a smart city is built on the development of ICTs like as big data, wireless communication, and the IoT. Things that were previously inconceivable in earlier cities are becoming possible as a result of the advancement of new technology. Digital devices and Internet networks are examples of smart technology, have been continuously studied, and a variety of inventions and services that were developed independently and subsequently linked together have been established [12]. Despite the fact that technology infrastructure is an important component of a smart city, its impact may be restricted if there is no human infrastructure in place. Even if a power plant is built to provide energy, it will be meaningless unless it is backed up by human infrastructure. Human infrastructure is as critical as technological infrastructure. This is why people must be educated on how to construct a smart city so that cutting-edge technologies can be utilized more effectively.

Smart digital infrastructure improves operational understanding and control, as well as the efficient use of scarce resources in a city. One of the primary benefits of ICT in a smart city is the capacity to record and distribute data in real time. Cities can take action before the situation worsens if data is delivered in real time and is reliable. Another way to think of digital infrastructure is as digital supporting layers (**Figure 2**).

- City: A meeting point for visual and digital infrastructure. Smart buildings, smart navigation, smart grids (for resources like water, electricity, and gas), and intelligent waste management systems are just a few examples.

- Sensor: This layer incorporates intelligent devices that measure and monitor variables in the boundaries of the city and its environment.

- Communication: This layer comprises the storing and transmission of data and information from the sensor level to data connections for further analysis.

- Data analysis: This layer includes analysis of data collected by different groups smart infrastructure systems, to help predict specific events (such as traffic congestion).

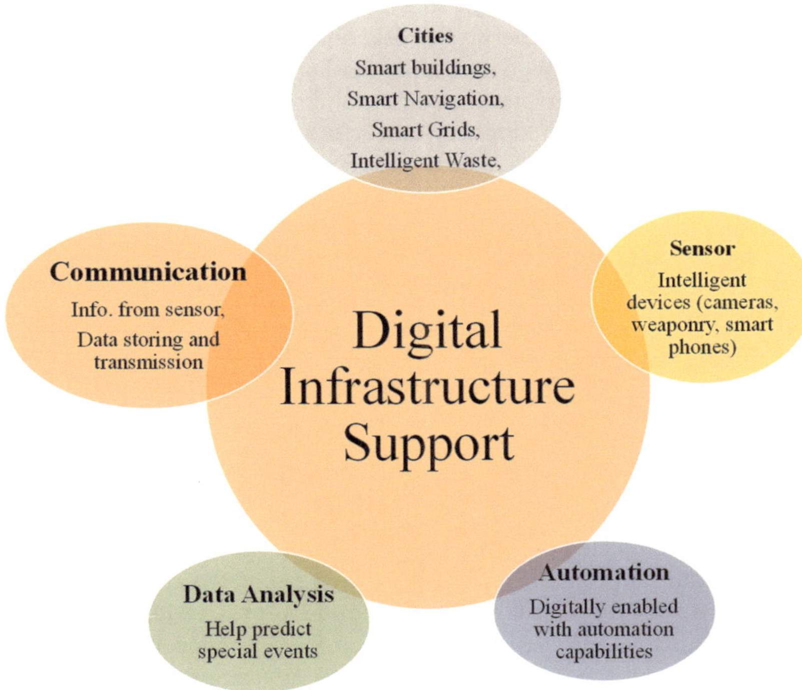

Figure 2.
Digital infrastructure.

- Automation:A digitally enabled interactive layer with automation capabilities. Vertical location and measurement of a huge number of devices in many areas

2.2 Smart city – collaboration system

Smart cities have a large influence on various facets of human life, including transportation, education, energy, and health. Weather information data, as an example, the amount of data on weather information is rapidly expanding. For agricultural development, identifying and extracting helpful information from the huge amount of weather data can be very beneficial. In addition, weather data analytics can help inform people ahead of time about potentially dangerous situations (e.g., extreme heat, flood information, drought, etc.) [13]. Governments have begun to adopt smart city concepts in order to improve residents' living conditions and execute big data applications [14]. Big data plays an important role in the smart city in order to change the economic condition of the country and its potential. By fulfilling the primary smart environment features, cities are able to realize the learning principles and requirements of the smart city applications. These characteristics include sustainability, resilience, governance, improved quality of life, and intelligent management of natural resources and municipal services, to name a few [15].

Beginning with independent operations, information sharing and communication are at the top of the hierarchy of relationships between governments, service providers, and people, followed by cooperation, coordination, collaboration, and eventually consolidation. The levels of relationships between governments, service providers, and people, are followed by cooperation, coordination, collaboration, and eventually consolidation, in which the organizations combine into a new, unified entity. The characteristics of collaboration that have been recognized by: Long-term relationship perspective, goal of achieving a previously unattainable outcome, substantial integration, synergy between organizations, systems have been altered, tight links between actors, actors move outside of traditional functional areas, possibly a new entity and highly interdependent, and power sharing [12].

Most big data applications for smart cities need intelligent networks that link many components, including citizens. Automobiles, smart home gadgets, and cellphones are examples of such devices. This network must be able to convey gathered data effectively from its source to the location where big data is collected, saved, and stored. The smart city processes the response and sends it to the various entities that need it. Quality of service (QoS) network support is critical for smart city real-time big data applications. In these applications, all current decentralized application events should be broadcast in real time to where they can be processed. These events are sent from the source as a raw event or as a filtered or aggregated event. Internet of Things (IoT) can be used to detect and collect multiple objects for the use of a smart city. Remote control requirements are usually implemented using the available network infrastructure. The flexible integration of many of the smart city's features will be opened as a result of this collaboration initiative [16].

2.3 Smart city – benefits

Cities are becoming more popular as a place to reside. As a result, there is an increasing demand for efficient urban management. In the case of mega-cities, this is especially true. The city and its residents will receive a lot of important benefits if they grow under the smart city model, which actively leverages IoT and other information technologies. The city and its residents will benefit from a number of great benefits:

- The city's utility efficiency has improved.
- Improved the city's traffic flow quality
- The living standards of city people have generally improved.
- City lighting systems efficient management
- Increased public transport convenience and efficiency
- Timely transport services and increased road safety
- All city services operation centralized monitoring
- Increase health improvement system and update timely
- Install the latest crime detection system

When comparing the past to the present, you can see that most cities have built full IoT ecosystems, which provide residents with several advantages such as mobility, security, health care, and enhanced efficiency. Every day, the environment changes, and we are confidently advancing toward the next industrial revolution.

3. Role of key technologies in smart cities

Smart towns are interconnected cities that employ anything from IoT sensors to open data collection and smart lighting to improve services and communication. Smart cities are no longer the wave of the future, they are the technology of the current age and growing rapidly because the Variety of software, IoT, Blockchain, Geographic Information System (GIS), AI and communication network are spreading around the world and its services are everywhere. These technologies are key to the growth of smart cities around the world. In smart cities data are collected through the citizens, buildings, infrastructure, assets, etc. and technology help us to use data to manage and monitor different systems such as transportation system, utilities, power plants, waste, water supply, information systems, schools, crime detection, libraries, community services, and hospitals.

The IoT is an interconnected network of devices that interact and share data. Home appliances, vehicles, and on-street sensors, to mention a few, are examples of this. To be useful, the huge volumes of data generated by a smart city must be analyzed quickly. Data acquired from these devices is stored in the cloud or on servers, allowing public and private sector efficiency to be enhanced, resulting in financial advantages and better human quality of life. A smart city is called a sustainable smart city they must meet the following criteria:

- Efficient and sustainable management system of natural resources for all citizens.
- All citizens and government, private agencies must be dedicated to the goal.
- Infrastructures and agencies must be technologically prepared to provide new solutions and services that will make citizens lives easier.
- Citizens who can live and work in the city and benefit from its resources.

That is the way which has been highlighted how smart cities give citizens a more efficient and high-quality living, and the way they use technology to reach these goals.

3.1 Impact of 5G in sustainable smart cities

Technology innovations and smart cities stand to benefit greatly from the capabilities of 5G technology. As a result, latency is minimized, and several devices may be connected at the same time with increased upload and download rates. In addition to technological advancements like 5G and smart cities, cultural shifts, economic constraints, and an aging population are all contributing to this next wave of generational change. It is being studied how 5G technology will alter the metropolis-based IoT vertical businesses, and therefore how 5G will just be the primary engine of that shift. A wide range of industries may benefit from 5G technology in smart cities, including energy, health care and manufacturing; media and entertainment; automotive; and public transportation [17].

For the purposes of 5G, the four dimensions of the value chain: generation, transmission, marketing, and consumption are all connected. Improved power management, less downtime, and lower operating costs may be achieved via 5G's use of distributed energy resources, advanced measurement systems, and more ensures effective communication into power-generating grids [18]. Using 5G's enormous bandwidth and ultra-low latency power, industries across the board will be able to spur economic growth and innovation by creating new subindustries, cutting costs, and raising the standard of goods and services they provide to their customers. Lawsuits affecting the usage of 5G mobile communications and how various sectors of the smart city will evolve or establish themselves are examined in [18]. Reference [18] also points out how ITS-5G and its consequences in a smart city may be realized via the use of automobile communications. All of these factors are examined from various angles to have a better understanding of ITS's long-term impact on society. Economic growth and innovation across sectors will be fueled by the 5G transition, based on high throughput, IoT, and low ultra-low response power, driven by modernization, cost reduction, and efficiency and quality of service. As a matter of fact, it will open the door to a new problem that is not just technical, but social and moral as well.

The goal of 5G is to improve public health, public safety, transportation, smart homes, and smart traffic systems, among other things. In fact, by 2020, the connection will be real-time due to the 5G data speed set to supply, and the latency level will be reduced to less than 2 milliseconds. If really want to maximize profits for people, then the resources and tools that can be used to supply 5G should be made available in a sustainable and environmentally friendly manner. If more efforts are made, the areas will also be affected. It has reached a very promising, yet important crossroads. The Smart City concept is a reality, and it is beginning to take shape in several European and international cities. Communities now have to support small cellular deployments to allow for 5th generation communication infrastructure so that Smart Cities can achieve their full potential and reap huge benefits. This wireless technology development will provide Smart Cities with their improved infrastructure. The next 10 years will see the breakthrough in various RAN technologies, bringing the world to a place almost new.

Vehicle to Vehicle (V2V), Vehicle to Pedestrian Communication (V2P), Vehicle to Infrastructure Communication (V2I), and inter-vehicle communications were briefly discussed by Mustakim. Then the next level is to look at 5G car networks and car network connectivity. The particularly focused on the role of unmanned aerial vehicles (UAV) in 5G connectivity, which can help create and sustain smart cities, such as deep-based UAV-based learning with the help of mm-wave and UAV space to increase capacity and extract data. The field of wireless technology in the IoT world of the smart city was explored by Wang et al. They also explained the reason for using IoT and 5G UAVs in smart cities in their report [19].

Before 5G can be extensively spread, the global market has to examine the effect of 5G on the environmental, society, and the economy. As a result, this is critical in terms of identifying and addressing risks and hazards. With the advent of new 5G technologies, the capabilities of mobile devices will unquestionably expand. Additional advancements will alter how technology interacts with our surroundings. The paradigm change from radio frequency to mmWaves, and even the new cell signaling, will enable mass manufacture and the usage of a huge number of devices. Energy harvesting, other energy sources, 5G green technology, large IoT sensors, smart meters, and life cycle monitoring are among the methods used to achieve the metrics for sustainability. IoT sensor deployment is one of the most important ways to achieve 5G network stability.

Virtual networks and various radio communications in smart devices are the foundations of 5G, which is based on virtual networks. New programs will be created as a consequence of greater spectrum utilization. Because to network technologies like software-defined networking (SDN) and network function virtualization (NFV), new services will be interconnected between all network components. Mobile Edge Computing (MEC) will assist decrease network latency since data cannot be obtained from a package, and offering the product will enable several networks to be displayed in the online system. Real-time data statistics may help you get there. This will assist to defend and maintain important health as a consequence of these networks' efforts [20].

With its low latency little less than 1 millisecond, 5G fiber optic cables are a key component of sustainable transportation systems Cars equipped with 5G networks can connect in a safe, healthy, and dependable manner. Car-to-network, car-to-walk, car-to-cloud, car-to-grid, and car-to-device interoperability modes may all be installed on the vehicle infrastructure. Using 4G, 5G, WiFi, and Bluetooth, vehicles may exchange knowledge/analysis about their speed and position. As a result of these innovations, drivers are better equipped to avoid collisions, boost traffic flow, and save fuel [21].

Secure data transmission for sophisticated analysis will be possible with 5G networks. Data plays a critical role in decreasing costs and enhancing efficiency in the healthcare business. 5G is predicted to have a latency of less than 1 millisecond, which will enable the edge computer to process data more quickly. In-home and outpatient surgical centers, walk-in clinics, care centers, and outpatient health care facilities may all benefit from the implementation of 5G in the medical industry. The ability of the hospital to transfer huge picture files may be improved as a result of this as well. It takes less time to transfer or receive data when a channel has a lot of bandwidth [22].

By creating an autonomous robot for inhabitants, the Smart Cities "Integrated Vision" aims to achieve its goal. In order for the Smart City environmentalist initiative to be successful, there has to be strong coordination between the many stakeholders. Today's 3G/4G wireless technology cannot depend to supply in-depth information necessary for the Smart City vision, such as the reliability of short delay and power efficiency of devices and more. It is for this reason that 5G is a Smart City goal providers network since it is essential for IoT, which is the backbone of Smart Cities. Network sensors and data will be used by emerging communities to offer municipal services more efficiently and effectively. Due to the enhanced IoT capabilities that 5G will provide, Smart Cities will flourish. 5G's ability to break down pricing barriers and open up hitherto untapped markets is critical if Smart Cities are to be a success. The development of 5G's network buildings and infrastructure enabling apps may bring up new employment prospects for smart cities [23].

3.2 Impact of 6G in sustainable smart cities

The sixth generation of wireless technology is known as 6G. Building on the reconstructed infrastructure and enhanced power currently being developed in the 5G millimeter-wave networks, the 6G network will follow in 4G and 5G [24]. It will provide networks with greater speed and reduced latitude through high frequency radio channels, allowing them to adopt sophisticated mobile devices and systems such as non-motorized vehicles [25].

The goal of 6G communications is to improve on the standard set by 5G communications by offering better network data availability, mobile data throughput, and seamless pervasive connectivity. In addition, 6G communications will use a revolutionary communication technique to get acceptance for a variety of mobile data categories and provide them over improved radio-frequency networks [26]. Many smart apps are integrated with 5G wireless communication technology. 5G standards, on the other hand, greatly highlight the need for new and emerging technologies. Data rate, volume, delays, reliability, resource sharing, and power/bit are some of these. To address these needs, the research focused on 6G wireless connectivity, enabling new technologies and applications [27].

6G offers very high data rates, up to 1 Tb/s, very high power, capable of supporting battery-free IoT devices; has control of low downtime delay has very wide frequency bands. Global broadband network streaming via global wireless integration with satellite systems; is a genius connected with the power of machine learning [28]. The structure of 6G networks should be designed to manage communication, processing, storage, and resource management as components of a cohesive system where its complete management requires efficient interaction. It will provide ICT infrastructure that allows end users to see themselves surrounded by a “great performance brain” that provides almost zero-latency services, unlimited storage, and great cognitive power [29]. Globally, the 6th generation (6G) mobile communication system is vigorously driven. Another way to achieve ultra-high-speed connectivity is to use terahertz bands above 100 GHz, which have a much wider frequency band than 5G. Dependence on frequency loss and channel features should be studied in order to detect 6G service frequency bands based on system performance [30].

The use of 6G is promoted as a way to improve automatic driving performance. Telephone driving (also known as slow-moving cars) is a concept when one is using a car at a distance. Deep-sea research and interplanetary have both used tele-operated driving. Tele-operation uses 5G networks tested by companies such as Ericsson and Huawei. Calling will require communication between the driver and the vehicle, especially when faced with an accident and the need for immediate response. If this is done correctly, it will improve future car rental services. It is also important to have a high level of security, privacy, and integrity of the network. Although research focuses on completely independent vehicles, telecommunications are desirable when autonomous mode fails or a complex situation requires human participation [31].

Several social pressures plaguing 4G wireless networks have been severely curtailed by 5G networks. Environmental protection and education, for example, have seen significant improvements in the 5G era. However, problems with connectivity and urbanization persist. 6G will no doubt provide a sense of relief. For a fully unlimited community, a hyper-connected user data connection will be full, and regional restrictions will be violated. Communication capabilities with many 6G features will contribute significantly to global sustainability and provide greater support for a variety of services in the application phase [32].

Nanotechnology, biotechnology, cognitive science, and ICT will all focus on 6G. Ultimately, this will raise public demands for sustainability, sustainability, openness, and inclusion, leading to complex social integration. In addition, 6G will promote productivity and rapid economic growth in rural and urban areas, helping to achieve sustainable goals [33]. The business environment will change drastically thanks to 6G. The changing business environment and the smooth and automated collection of market data from individuals will determine the future of the business. With state-of-the-art products and specialized services, 6G will provide an easy-to-use platform for intelligent data processing. These products and services will be designed to be extremely sustainable and customized to meet the unique needs of consumers in rural and urban areas. In addition, 6G will allow for the mobilization of people and the development of high-quality distribution platforms to promote the sharing of sustainable business models and accelerate equitable distribution of resources. Recently, options for establishing a world-class, long-term 6G business future for all future 6G business partners were re-evaluated [34].

3.3 Impact of AI in sustainable smart cities

In a smart city, one of the most crucial tools is AI. It is a sub-field of computer science that focuses on enhancing machines' cognitive skills and creating artificially intelligent beings. Searches, mathematical computations, logic, algorithms, and Bayesian and economical procedures are among the many instruments available. There are several definitions of smart cities offered by various academics. To become a smart city, however, a city must use ICT and AI to accomplish long-term social, environment, and economic development while also raising the quality of life for its citizens. A technologically interconnected city or the use of AI technology with big data to produce intelligence & efficiency in managing the city's resources [35] might be used to describe the technology component of a smart city.

In a study [10] on good decision in smart cities using big data, researchers established a three-layer framework that describes a smart city as instrumented, networked, and intelligent. AI and IoT are used to gather real-time data from surveillance cameras, monitors, and sensor-based devices as well as from open data sources and social networking sites for rapid reaction in the implementation phase of smart cities. In the "interconnected" phase, data from AI, IoT, and other sources is combined and transformed into a piece of relative knowledge to provide greater insights for smart decision-making. Finally, the city's demands, requirements, needs, and policies will be understood using converted data obtained through data. As a result, it can help people make educated and wise decisions [10].

In a Smart City project, the essential infrastructure elements which include AI are:

1. Public transportation system management.

2. Reliable electricity supply.

3. Cleaning and waste management.

4. Public transportation system management.

Also, e-governance and citizen collaboration will be used to attain these objectives. This contains ambitious plans such as:

1. Service delivery using electronic means.

2. Citizen – the city's eyes and ears.

3. Public information and redress of claims.

4. Video – Criminal Surveillance.

AI in smart cities will play a key role in making urbanization smarter so that it is a sustainable growth that makes cities armed with improved aspects of life, hiking, shopping and enjoying a safer and more appropriate life in such an environment.

In fact, while developing smart cities, a few difficulties such as management, sanitation, traffic congestion, security monitoring, parking management, and much more may be addressed with AI to give a long-term solution for residents. Smart Mobility solutions aim to increase safety and efficiency, reduce traffic congestion, improve air quality and noise pollution and cost reduction. Wise mobility solutions are also seen as important for moving forward to decarbonize the transport sector and achieve EU pollution reduction targets. AI is powerful an emerging tool that prides itself on the ability to drive sustainable change into efficient, sustainable resources and person-centered travel plans, especially in urban situations.

In short, urban planning is about solving the challenges of modern civilization. These problems are compounded by the growth of modern society. Problems in the community range from general to technological, such as ensuring sanitation and management of infrastructure. Wise cities have recently aroused the curiosity of social scientists, engineers, and anyone interested in incorporating technology into their daily lives [36]. AI and IoT have become an integral part of our daily lives. With the proliferation of smart gadgets connected to the Internet, data is everywhere. Smart solutions for smart cities can be built using this data. The effects of AI and IoT on urban life are encouraging. Because of its amazing ability to transform everything, AI is sometimes called the fourth industrial revolution. AI blesses humanity intelligent health care to secure intelligent cities as it evolves day by day [37].

AI creates a modern world. From nineteenth century, planning of cities has focused on improving the quality of life of the people by focusing on the economic functioning of cities and social justice. Many tasks are common and are done mechanically and humanly, but some systems are still working that require the involvement of a sensitive human mind. Many common tasks are expected to be replaced by clever operating principles in the near future, but those that require significant skills, such as design, will take longer to automate [17].

The present state of AI necessitates the storage of enormous volumes of data. Data about how people actually live in cities and just how cities change over time may be accessible due to the rise of closed-circuit television, sensors, and humongous communication networks in contemporary cities. Other objects such as land, buildings, and open areas may be accessed. Using this data in an intelligent system, such as machine learning models, urban planners may utilize it with precision understanding and design to learn about the city's fabric. It is possible to build an urban government using information gleaned from numerous digital sources. Politicians may be able to design better policies for city dwellers with the use of data from an analytical model.

As climate change is a big worry in today's world, AI systems has also been used to produce city planning strategies that lessen the consequences of climate change & policies that assist mitigate climate change. There are also concerns about privacy breaches when it comes to AI-based urban planning, which is a human activity in the future. The privacy of residents living in smart cities is severely compromised as large data must be collected and stored in order to create an AI model. Because data must be stored locally, cyber criminals may try to access data centers and gain illegal access. As a result, regulations and strategies to protect human data are needed [7].

A variety of AI-based applications are designed for intelligent and sustainable cities. They have introduced a method that uses a calculator lens with a small microscope and machine learning methods to determine air quality. Their solution is called C-Air, and includes a smartphone app that can manage and display multiple settings and findings. It contains a machine that can take small pictures of particles in the air, as well as a machine learning model that can predict what particles are in the image and their size. They have used a machine learning algorithm they have made for themselves [38]. They have introduced a Smart Traffic Management Platform that can use big data and smart algorithms to improve traffic flow [37].

3.4 Impact of block chain in sustainable smart cities

Blockchain is an open and shared ledger technology (DLT) that can successfully record, permanently, and securely record transactions between two parties. Integrates a system with a distributed network of digital data that is copied and synchronized across multiple devices. The main goal of the DLT is to build trust, accountability, and transparency without relying on a single source of authority or in situations where players do not trust each other. It also improves data integrity and geographical distribution. The advent of Blockchain technology as a transparent and responsible way to protect data opens the way for complex data privacy, security, and integrity concerns to be resolved within the smart city ecosystem. Applications that include data access, control, and distribution of patient records, electronics, and financial management. Advanced technology and state-of-the-art networks are important drivers of urban efficiency in a smart city.

Key components of a smart city ecosystem, such as its major infrastructure and e-government services, are interacting in real time. Wireless communication networks, combined with self-planning and livelihood networks, are essential to the development of smart cities. By delivering critical security services to ensure authenticity, confidentiality, integrity, and accessibility, high-speed, real-time security agreements are an integral part of the smart city ecosystem [39]. Blockchain technology has the ability to rebuild intelligent city infrastructure, change ecosystems to access improved consumer services, and enable new applications. Because it enhances efficiency, protects sensitive data exchange, and enhances smart city systems integration, the Blockchain is revered as a new development and wealth engine in the smart city. Researchers and experts believe that, in addition to supporting cryptocurrencies, Blockchain technology could help re-establish urban development around the world by acquiring transactions with other services [40].

It can be thought of increasing traffic stability by reducing energy consumption, improving safety, and reducing pollution through Blockchain-based Internet of Vehicles (IoV). The introduction of Blockchain, as well as the transition from fuel-efficient vehicles to electric and independent Blockchain powered vehicles, has the potential to create new business models where mobility as a service replaces conventional car ownership ideas. Citizens will benefit from a high level of smart travel, real-time public transportation tracking, fast payment services, ample parking, and easy walking as a result [41].

Physicians can store patient health data in a Blockchain that can be set up to allow interaction between different healthcare companies. In addition, Blockchain programs can provide real-time access to patient medical records while providing protection against data fraud that is difficult to track, such as adding or deleting drug allergies, critical patient safety and institutional trust concerns. The fact that the Blockchain enables security, privacy, and integrity of data without the need for an outside company to control the activity arouses interest in technology. For smart city administrators, ensuring that security becomes a priority to combat cybercrime. To address the security and privacy challenges facing smart cities, the Blockchain must be integrated to ensure that certain security errors do not continue to affect all other smart city networks [42].

3.5 Impact of geographic information system (GIS) in sustainable smart cities

Smart cities are defined as areas of big cities that thrive on sustainability and provide extraordinary living conditions by improving economic, environmental, transportation, governance, and energy efficiency, among other things. Smart cities use a network of sensors, cameras, wireless tools, and data centers to enable integrated city monitoring and administration through the use of technology like GIS, Global Positioning Systems (GPS), and remote sensors (RS). It is essential for transforming a city into an intelligent city. GIS is used in surveying, engineering, and organizing the collection, processing, management and presentation of location data, in addition to the map display [43].

High-resolution satellite imagery helps to prepare land use maps of cities showing agricultural land use, residential, industrial, commercial, social, and low land use. Parts of a smart city include smart planning, public administration, smart power, smart buildings, smart infrastructure, public safety, smart security, smart traffic management, smart waste disposal and smart service delivery methods. GIS unifies all parts of city planning and management, providing a one-stop shop for everyone. Because smart cities include fewer participants, project success depends on the integration, networking and collaboration of various actors in the smart city ecosystem [44].

A smart city is one in which investments in human and social capital, traditional and modern transportation, infrastructure, and long-term economic growth result in a greater standard of life by collaborating on natural resource management. A city with the necessary infrastructure to provide a clean and sustainable environment through new solutions [45]. Smart Solutions will allow communities to improve infrastructure and services by integrating technology, information, and data. The need for a single technology platform to facilitate the integration, integration, and collaboration of various actors in the smart city ecosystem is part of the key achievement. GIS can play an important role in establishing government-citizen interactions where citizens can communicate with concerns, make feedback on local infrastructure, and learn about measures to improve the city [46].

The city of the future will no doubt be very different from the city of today, and none of us can fully predict the changes that will take place. The connected and private vehicles will present more real-time GIS opportunities; that major changes are taking place in our retail systems as a result of switching to online services; that new digital technologies are changing the diversity of economic markets; and that climate change will affect many aspects of urban life, especially in parts of the world where food, energy, and water are nearing critical levels. Our future will be dominated by global population change and great success in health care, all of which will bring endless opportunities [35].

GIS systems have become an integral part of city planning. GIS has been implemented in the information infrastructure of all major cities. It is used in the investment process and strong management of well-known projects. Although GIS is not a new solution, new applications have emerged in recent years. The Internet of Things and other tools and technologies connected to the Smart City concept are booming. As a result, new potential areas for data collaboration and resource information and integration are developing [17].

3.6 Impact of IoT in sustainable smart cities

IoT technological advancements and its implementation into intelligent cities have transformed our work and living environments, while enhancing our civilization. There are a number of downsides to IoT technology in smart cities, including higher energy consumption, dangerous pollution levels, and the generation of electronic garbage (E-waste). Green IoT is required for smart city applications to be ecologically sustainable. As a consequence of green IoT, smart cities are more environmentally friendly, which makes them more sustainable. As a result, it is important to address the threat of pollution, traffic congestion, resource use, energy consumption, public safety, quality of life, environmental sustainability, and cost management strategies and strategies [47].

The objects around us are integrated into many intelligent city applications, enhancing our quality of life, thanks to dramatic advances in communication and sensory technology. Internet of Things is a term used to describe the interaction of objects in a smart city. In smart cities, IoT refers to everything that can be connected anytime, anywhere, to any channel [48].

IoT technology is developing rapidly, allowing IoT components to be intelligent through flexible communication network, processing, analysis, and storage, cameras, sensors, Radio Frequency Identification (RFID), actuators, drones, cell phones, and other IoT devices are examples. All of these devices have the ability to communicate and work together to achieve common goals. IoT devices will be able to provide many real-time monitoring applications using such components and communication technologies, as evidenced by environmental monitoring, healthcare, transportation independence, digital industry and automation, and home automation. In addition, IoT allows software Agents to exchange information, make collaborative decisions, and complete tasks more efficiently [49].

IoT has a profound impact on smart cities, with its many programs affecting social transformation, reducing traffic congestion, creating less expensive municipal services, keeping citizens safe and healthy, reducing energy consumption, improving monitoring systems, and reducing pollution in various ways. However, scholars are focused on the natural challenges of IoT such as energy consumption, carbon emissions, energy conservation, trade, carbon labeling, and footprint [50].

A data center, on the other hand, is required for data management and conversion into smart city information, which would otherwise be impossible. As a result, it consumes a large amount of electricity, is expensive to run, and has a significant carbon impact. Many common gadgets, including handheld phones, actuators, sensors, and RFIDs [51]. contribute to the production of big data. In order to be considered "smart," a city must have a high standard of living, good environmental management, and a healthy economy. Power and water supplies, internet connection, smart parking, and other necessities for smart city applications should all be available in smart cities. Unlimited internet computer services and storage are made possible by cloud computing. An array of devices is shown to be linked together to collect data over the internet cloud. It is possible to create a complete learning environment using cloud computing and the IoT. A primary purpose of cloud applications is to encourage ecologically friendly products that may be readily reused and recycled [52].

Garbage collection and intelligent city planning must be implemented to provide a clean environment. The smart IoT devices, edge information, and cloud are being sought by businesses and governments alike as a low-cost means to gather various forms of rubbish. It is necessary to build, run, and improve an automated trash collecting system in order to get the most out of its use, storage, and production capabilities. By permitting real-time supervision and cloud connection, the IoT may assist enhance automated trash collection systems. In addition, the authors stress the need of automating trash collection systems in order to increase productivity and efficiency. They examine how smart city infrastructure may be integrated with technology. Real-time supervision and data gathering have been made possible thanks to the IoT [53].

To send and receive data, a smart city is built on an intelligent foundation and a complex system of ubiquitous networks, objects, government, and communications. The data collected in the cloud of smart cities of any app is handled and analyzed appropriately, allowing decision-making based on available facts and real-time action to improve the way we work and live. Research is investigating the importance of smart cities in creating sustainable cities. Air quality, renewable energy, energy efficiency, water quality, and environmental monitoring are all major concerns [37].

Green IoT is important for smart cities to create environmentally friendly and sustainable places to work and live. Raw IoT strategies and technologies surpass traditional IoT strategies and technologies in big data analysis, making smart cities safer, smarter, and more stable. The authors examined how big data has improved living standards by reducing land pollution and using resources efficiently [54].

In the field of smart cities, IoT devices and technologies have gained a lot of interest. It is necessary to address the definition of waste management and smart communities. Smart cities propose the use of a large number of smart devices capable of processing and computing to facilitate green automation, monitoring, and data collection. Understanding the field of waste management and determining value in controlled waste collection and disposal requires understanding the permissive, planning, social, and economic aspects. In addition, the efficient management and collection of wasteful city infrastructure should be considered. The link between waste management and the activities of smart communities must be handled in a consistent manner [7].

IoT has the potential to transform the healthcare industry by shifting its focus from therapies to ensuring the well-being of everyone. This field, however, is still in its infancy, and a few things need to be investigated before its full potential can be determined. Discusses the effectiveness of IoT in preventive health care and treatment in relation to a variety of workplaces such as disease monitoring, age-based monitoring, body abnormalities, and profile-based monitoring; and presents open-ended research questions and future research guides on IoT use [39].

Smart homes, workplaces, schools, data centers, industries, and warehouses are examples of smart infrastructure. IoT technology can be used to control security, surveillance, automation, power management, and other advanced architectural features. Smart homes, workplaces, warehouses, and other smart buildings complete their obligations quickly and efficiently [40].

Smart appliances and systems use sensors to monitor the environment and take appropriate action, such as flashing lights or turning off the air heater. These clever systems also help in predicting demand. Similarly, smart warehouses can help improve supply chain management production. The biggest advantage of smart houses and buildings is the convenience they offer users, as they are released to focus on other activities. Intelligent health care systems need to be in place for cities to thrive [41]. IoT applications have the potential to significantly improve urban transport infrastructure.

Among the features of an intelligent urban transportation system are the automatic identification numbers, road vehicle counts, traffic signal automation, intelligent lighting, and intelligent parking. The use of Internet of Things to manage vehicle traffic information can help control real-time traffic, benefiting citizens, city governments, and the urban environment. The combination of sensory capabilities, modern GPS-enabled vehicles, air quality, and sound sensors used on a particular road can greatly assist in traffic monitoring and city sustainability [42].

Residents are concerned about the lack of adequate parking in cities. The Smart parking lot software can track the number of vehicles in various parking lots throughout the city, as well as their arrival and departure. Drivers may use street sensors and sophisticated displays to determine the correct parking route in the city. Users, sellers/contractors of parking spaces, government, and the general public all make a profit in smart parking spaces. Finding a parking space quickly reduces traffic congestion, less pollution, and happier residents [46].

4. State-of-the-art

Global urbanization is moving quicker than ever before, and it is happening all across the planet. Global urbanization peaked in 2007 at 51%, and it is expected to reach 70% by 2050. Around 60% of the world's population is expected to be living in cities by 2050, up from the current 1.4 billion urban residents in 1970 [55]. Fresh urban issues are prompting new debates about how to deal with them. One and among the most sought-after solutions is the creation of smart cities. An urban development strategy known as the smart city involves constructing cities with the use of ICT. Cities are huge cause of pollution, congestion, for waste, but they also exacerbate a variety of socioeconomic concerns, such as increased poverty, crime, and unemployment, because of unchecked population and growing resource demand, combined with poor organization and management. Because of this, urban management is one of the most serious concerns of the twenty-first century, necessitating innovative measures in industries such as engineering, public safety, and the natural environment [56].

In a smart city environment, The Internet of Things provides framework for connecting gadget allowing for easier information transmission across platforms [49]. As a result of the recent adoption of a variety of wireless technologies, like as IoT is ready to become the next disruptive technology that takes full advantage of the potential of the Internet. Smart retail, energy, transit, water, housing, healthcare and grids all recently observed IoT being used to build intelligent systems in smart cities.

However, There is no globally accepted definition of a smart city, and it is difficult to spot shared global patterns [49]. The concept focuses on integrating next generation information technology into all parts of life, including hospitals, electrical grids, railroads, bridges, tunnels, highways, water systems, buildings, oil, dams and gas pipelines, and other items all over the world [57]. The Internet revolution resulted in unparalleled levels of connectedness and speed among individuals. The connectivity of items to build a smart city will be the next revolution. The interconnectedness of sensing and actuating equipment is emphasized in the smart city, a standard framework that allows information to be transferred across platforms. Such sharing is enabled by cloud computing, which serves as the unifying foundation for data analytics, omnipresent sensing and information representation. The post PC era has arrived, and smartphones and different hand-held devices area unit reworking the environment by creating it additional interactive and informative [58].

The Internet revolution resulted in unparalleled levels of connectedness and speed among individuals. The connectivity of items to build a smart city will be the next revolution. The interconnectedness of sensing and actuating equipment is emphasized in the smart city, a standard framework that allows information to be transferred across platforms. Such sharing is enabled by cloud computing, which serves as the unifying foundation for data analytics, omnipresent sensing and information representation. The post-PC era has arrived, and smartphones and different hand-held devices area unit reworking the environment by creating it additional interactive and informative [49].

Big data systems have been successfully stored, analyzed, and extracted in smart cities to provide information that will be used to improve a variety of smart city services. Big data could also help planners anticipate future expansion in smart city resources, services, or places. The multiple aspects of big data emphasize its significant benefits and improvements potential. The possibilities are limitless; nevertheless, the availability of modern technology and equipment limits them. With the proper tools and methodologies for economical and effective knowledge analysis, huge knowledge will fulfill its aims and develop sensible town services. Such efficiency would encourage collaboration and communication among organizations, as well as the development of new services and apps that will strengthen the smart city.

Big data applications may benefit a wide range of industries in a smart city, resulting in improved consumer experiences and services, as well as improved corporate performance. Diagnostic and treatment tools, Preventive care services, healthcare records administration, and patient care may all be improved (see **Figure 3**). Big data may help transportation networks optimize routes and timetables, handle fluctuating needs, and improve environmental friendliness.

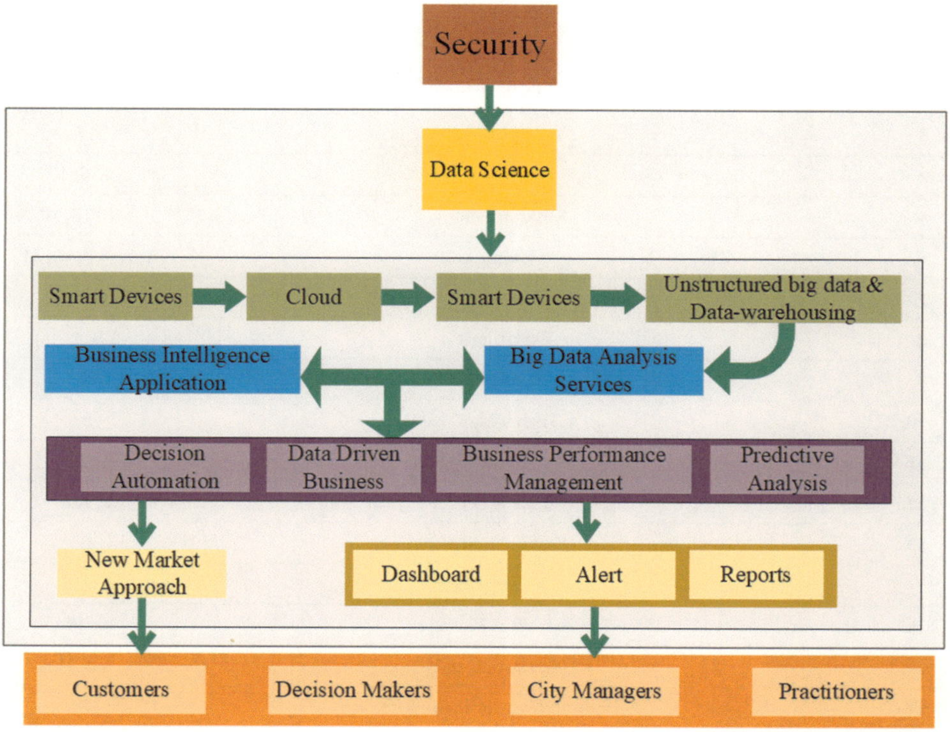

Figure 3.
Smart city and Business model for big data.

The smart city age of big data has ushered in a lot of new value-creation possibilities, which faces several difficulties, the majority of which are multi-dimensional and may be tackled from various interdisciplinary viewpoints [57].

Cloud computing refers to a set of computing models that include a large number of machines or clusters linked by a network system. It allows users for executing complicated computing operations on vast area like mining enormous amounts of social network data provided by smartphone apps [57]. The basic engine for cloud computing is provided by big data technology such as the Hadoop framework. Hadoop was intended to provide a platform and programming paradigms for distributing big dataset processing across several clusters. Hadoop is made up of two main components: Hadoop Distributed File System and MapReduce, both of which are intertwined [59], although the smart city's real-time data storage and processing requirements are taken into account. Using streaming architecture, the network's sensing devices will be able to communicate efficiently and smoothly.

4.1 Smart city performance evolution

The smart city diagnostic model's indexes, such as Environment, Living, Traffic, Governance, and Plan/Strategy have been observed to have a high frequency of smart city performance evaluation indices, see **Table 1**. The smart city diagnostic model's indexes, such as Environment, Living, Traffic, Governance, and Plan/Strategy were shown to have a high frequency of smart city performance evaluation indices. There are unit limits in this it is troublesome to spot good cities from property indicators in

Sr. No.	Category	Frequency
1.	Economy	1, 2, 8, 10, 11
2.	Human Resources	1, 2
3.	Governance	1, 2, 6, 8
4.	Traffic	1, 2, 8, 10
5.	Environment	1, 2,7, 8, 10
6.	Living	1, 2, 7, 8, 10, 11
7.	Strategy	3, 4, 5, 14
8.	Execution	3, 4
9.	Culture	5
10.	Technology	5
11.	Data	5, 10
12.	Innovation	6, 7, 8
13.	Sustainability	6
14.	Social Integration	6
15.	Information Linkage	6, 10
16.	Infrastructure	7, 8, 11
17.	Crime and Disaster Prevention	7

Table 1.
Smart city performance evaluation index frequency [60].

key EU-based diagnostic models, additionally as international cities and inexperienced town indexes, as a result of good cities area unit recognized as technology and systems additionally as human parts.

When it comes to the smart city diagnosis model, there is a proclivity to focus on European technology, systems, and human aspects. Most domestic diagnostic models, on the other hand, have narrowly concentrated on technology elements that operate as infrastructure construction. The Navigant index, as well as the Global System for Mobile Association (GSMA) and Erickson indices, which are mental in nature and evaluate private enterprises, have limitations due to their overemphasis on mobile functions. Of course, technological infrastructure plays a major role within the development of a sensible city, and it is integrated with ICT.

However, the above-mentioned characteristics cannot be used to draw judgments about a good smart city. In order for a city to develop, many factors must be developed in consideration of the internal and external environments, including leadership, organizational structure, governmental system, legal backgrounds, political processes, interest groups, citizen support and participation, local industries and vendor communities, and stakeholders. Look at how non-technical concerns like people's collaboration, government policy support, leadership, and local innovation should all be included in smart city design and growth.

5. Conclusions

A second historical wave of migration from rural to urban areas is currently beginning. According to the most recent United Nations study on Smart Cities, the world would wish to develop 10,000 new cities by 2040, where China has already committed to making 100 new cities to accommodate the 385 million people that are expected to migrate from the rural area to the town. There will be seven new cities in Korea, six in Asian countries, and personal sector initiatives like PlanIT vale in Portuguese Republic and Lavasa in Asian country square measure within the works. Utilization way for Europe states that smart use and exploitation of technologies and knowledge will help us face the challenges confronting society and Europe. There is a clear trend towards increasing the percentage of people living in cities in the near future. These urban conglomerates must therefore handle and resolve the bulk of society's issues, sometimes with scarce resources and sophisticated progressive groups, in which judgment call becomes a cumbersome and inefficient process with a lack of openness.

Many academics from numerous fields are interested in the fast growth of data because of the large growth in connected devices in urban locations. This chapter, therefore, highlights the role and importance of emerging information and communication technologies in new generation of sustainable smart cities. The elements of sustainable smart cities have also been highlighted by giving an overview of how cities throughout the world have adopted them and projected trends for the next generation of sustainable smart cities. Section 2 discusses about the importance and need of smart cities in our day to day life. The various dimensions that need to be considered while looking at the smart cities are depicted using **Figure 1**. Section 2.1 provides infrastructural requirements for smart cities. Section 2.2 explicates that smart cities project success depends on the integration, networking and collaboration of various actors in the smart city ecosystem, while Section 2.3 highlights how the cities and its residents will benefit from a number of great benefits in smart cities. Section 3 provides Role of key technologies in smart cities, for instance, 5G, 6G, AI, Blockchain and IoT. Section 4 provides the state of the art in terms of the steps towards future sustainable smart cities and finally Section 4.1 summarizes the discussion by providing the smart city diagnostic model indexes, such as environment, living, traffic, governance, and plan/ strategy in **Table 1**.

Author details

Kamal Shahid*, Muhammad Hassan, Ali Husnain and Sadaf Mukhtar
Institute of Electrical, Electronics and Computer Engineering, University of the Punjab, Lahore, Pakistan

*Address all correspondence to: kamal.ee@pu.edu.pk

References

[1] Bhatnagar S, Garg D, Bhatnagar M. Smart cities – An overview and the role of ICT. Research Cell: An International Journal of Engineering Sciences. 2014;**13**: 1-5

[2] Börjesson Rivera M, Eriksson E, Wangel J. ICT practices in smart sustainable cities: In the intersection of technological solutions and practices of everyday life. EnviroInfo and ICT4S. 2015. pp. 317-324

[3] Tcholtchev N, Schieferdecker I. Sustainable and reliable information and communication technology for resilient smart cities. Smart Cities. 2021;**4**:156-176

[4] Shehab MJ, Kassem I, Kutty AA, Kucukvar M, Onat N, Khattab T. 5G networks towards smart and sustainable cities: A review of recent developments, applications and future perspectives. IEEE Access. 2022;**10**:2987-3006. DOI: 10.1109/ACCESS.2021.3139436

[5] Al-Mshahde M. Reliability-Oriented Intra-Frequency Dual Connectivity for 5G systems: Configuration Algorithms and Performance Evaluation. Aalborg, Denmark: Universitat Politècnica de Catalunya; 2018

[6] Ghorbani, H., Mohammadzadeh, M. and Ahmadzadegan, M. Modeling for malicious traffic detection in 6G next generation networks. In 2020 International Conference On Technology And Entrepreneurship-Virtual (ICTE-V). pp. 1-6 (2020)

[7] Almalki F, Alsamhi S, Sahal R, Hassan J, Hawbani A, Rajput N, et al. Green IoT for eco-friendly and sustainable smart cities: Future directions and opportunities. Mobile Networks and Applications. 2021:1-25

[8] Bibri S. The IoT for smart sustainable cities of the future: An analytical framework for sensor-based big data applications for environmental sustainability. Sustainable Cities and Society. 2018;**38**:230-253

[9] Sharma P, Jangirala S. Internet of Things for sustainable urbanism. Journal of Physics: Conference Series. 2022;**2236**: 012008

[10] Ferraz, F. and Ferraz, C. Smart city security issues: Depicting information security issues in the role of an urban environment. In 2014 IEEE/ACM 7th International Conference On Utility And Cloud Computing. pp. 842-847 (2014)

[11] Anthony Jnr B. A case-based reasoning recommender system for sustainable smart city development. AI & Society. 2021;**36**:159-183

[12] Mills D, Izadgoshasb I, Pudney S. Smart city collaboration: A review and an agenda for establishing sustainable collaboration. Sustainability. 2021;**13**: 9189

[13] Fan W, Bifet A. Mining big data: Current status, and forecast to the future. ACM SIGKDD Explorations Newsletter. 2013;**14**:1-5

[14] Jiménez C, Solanas A, Falcone F. E-government interoperability: Linking open and smart government. Computer. 2014;**47**:22-24

[15] Al Nuaimi E, Al Neyadi H, Mohamed N, Al-Jaroodi J. Applications of big data to smart cities. Journal of Internet Services and Applications. 2015;**6**:1-15

[16] Atzori L, Iera A, Morabito G. The internet of things: A survey. Computer Networks. 2010;**54**:2787-2805

[17] Jha, A., Ghimire, A., Thapa, S., Jha, A. and Raj, R. A review of AI for urban planning: Towards building sustainable smart cities. In 2021 6th International Conference On Inventive Computation Technologies (ICICT). pp. 937-944 (2021)

[18] Soliman W. The economic function of network economy: A case study of North America. International Journal of Science and Research (IJSR). Jan 2022; **11**(1). DOI: 10.21275/SR22107175649

[19] Parcu P, Innocenti N, Carrozza C. Ubiquitous technologies and 5G development. Who is leading the race? Telecommunications Policy. 2022;**46**: 102277

[20] Alsharif M, Kelechi A, Albreem M, Chaudhry S, Zia M, Kim S. Sixth generation (6G) wireless networks: Vision, research activities, challenges and potential solutions. Symmetry. 2020;**12**:676

[21] Rao S, Prasad R. Impact of 5G technologies on smart city implementation. Wireless Personal Communications. 2018;**100**:161-176

[22] Ndiaye, M., Saley, A., Niane, K. and Raimy, A. Future 6G communication networks: Typical IoT network topology and terahertz frequency challenges and research issues. In 2022 2nd International Conference On Innovative Research In Applied Science, Engineering And Technology (IRASET). pp. 1-5 (2022)

[23] Chaudhary P, Gupta B, Singh A. XSS Armor: Constructing XSS defensive framework for preserving big data privacy in internet-of-things (IoT) networks. Journal of Circuits, Systems and Computers. 2022:2250222

[24] Elmeadawy, S. and Shubair, R. 6G wireless communications: Future technologies and research challenges. In 2019 International Conference On Electrical And Computing Technologies And Applications (ICECTA). pp. 1-5 (2019)

[25] Oinas-Kukkonen H, Karppinen P, Kekkonen M. 5g and 6g broadband cellular network technologies as enablers of new avenues for behavioral influence with examples from reduced rural-urban digital divide. Urban Science. 2021;**5**:60

[26] Verma A, Jaiswal R, Mishra S. An intelligence system for OFDM with subcarrier power modulation and STBC to improve system performance. Journal of University of Shanghai for Science and Technology. Jul 2021;**23**(7)

[27] Wang M, Lin Y, Tian Q, Si G. Transfer learning promotes 6G wireless communications: Recent advances and future challenges. IEEE Transactions on Reliability. 2021;**70**:790-807

[28] Letaief K, Chen W, Shi Y, Zhang J, Zhang Y. The roadmap to 6G: AI empowered wireless networks. IEEE Communications Magazine. 2019;**57**: 84-90

[29] Strinati E, Barbarossa S, Gonzalez-Jimenez J, Ktenas D, Cassiau N, Maret L, et al. 6G: The next frontier: From holographic messaging to artificial intelligence using subterahertz and visible light communication. IEEE Vehicular Technology Magazine. 2019; **14**:42-50

[30] Inomata, M., Yamada, W., Kuno, N., Sasaki, M., Kitao, K., Nakamura, M., Ishikawa, H. and Oda, Y. Terahertz propagation characteristics for 6G mobile communication systems. In 2021 15th European Conference On Antennas And Propagation (EuCAP). pp. 1-5 (2021)

[31] Imoize A, Adedeji O, Tandiya N, Shetty S. 6G enabled smart infrastructure for sustainable society: Opportunities, challenges, and research roadmap. Sensors. 2021;**21**:1709

[32] Bhat J, Alqahtani S. 6G ecosystem: Current status and future perspective. IEEE Access. 2021;**9**:43134-43167

[33] Cruz C, Sarmento J. "Mobility as a service" platforms: A critical path towards increasing the sustainability of transportation systems. Sustainability. 2020;**12**:6368

[34] Yrjölä S, Ahokangas P, Matinmikko-Blue M. Sustainability as a challenge and driver for novel ecosystemic 6G business scenarios. Sustainability. 2020;**12**:8951

[35] Thapa, S., Adhikari, S., Ghimire, A. and Aditya, A. Feature selection based twin-support vector machine for the diagnosis of Parkinson's disease. In 2020 IEEE 8th R10 Humanitarian Technology Conference (R10-HTC). pp. 1-6 (2020)

[36] Singh S, Jeong Y, Park J. A deep learning-based IoT-oriented infrastructure for secure smart city. Sustainable Cities and Society. 2020;**60**: 102252

[37] Nallaperuma D, Nawaratne R, Bandaragoda T, Adikari A, Nguyen S, Kempitiya T, et al. Online incremental machine learning platform for big data-driven smart traffic management. IEEE Transactions on Intelligent Transportation Systems. 2019;**20**:4679-4690

[38] Nandanwar, H. Smart IoT based air quality monitoring systems for smart cities: Designs issues, implementation, analytics, and challenges. (PhD Dissertation). 2022

[39] Mehta R, Khanna K, Sahni J. IoT in Healthcare. In: IoT for Sustainable Smart Cities and Society. Internet of Things. Springer International Publishing; 2022. pp. 85-106

[40] Roman R, Zhou J, Lopez J. On the features and challenges of security and privacy in distributed internet of things. Computer Networks. 2013;**57**:2266-2279

[41] Jabbar S, Khan M, Silva B, Han K. A REST-based industrial web of things' framework for smart warehousing. The Journal of Supercomputing. 2018;**74**: 4419-4433

[42] Li X, Shu W, Li M, Huang H, Luo P, Wu M. Performance evaluation of vehicle-based mobile sensor networks for traffic monitoring. IEEE Transactions on Vehicular Technology. 2008;**58**:1647-1653

[43] Wahed, M. GIS for building smart cities. In 1st International Conference On Towards A Better Quality Of Life. (2017)

[44] Li W, Batty M, Goodchild M. Real-time GIS for smart cities. International Journal of Geographical Information Science. 2020;**34**:311-324

[45] Turek T, Stepniak C. Areas of integration of GIS technology and smart city tools. Research findings. Procedia Computer Science. 2021;**192**:4681-4690

[46] Lee, S., Yoon, D. and Ghosh, A. Intelligent parking lot application using wireless sensor networks. In 2008 International Symposium On Collaborative Technologies And Systems. pp. 48-57 (2008)

[47] Batty M. Artificial intelligence and smart cities. Environment and Planning B: Urban Analytics and City Science. 2018;**45**:3-6

[48] Allam Z, Dhunny Z. On big data, artificial intelligence and smart cities. Cities. 2019;**89**:80-91

[49] Gubbi J, Buyya R, Marusic S, Palaniswami M. Internet of Things (IoT): A vision, architectural elements, and future directions. Future Generation Computer Systems. 2013;**29**:1645-1660

[50] Arshad R, Zahoor S, Shah M, Wahid A, Yu H. Green IoT: An investigation on energy saving practices for 2020 and beyond. IEEE Access. 2017;**5**:15667-15681

[51] Talari S, Shafie-Khah M, Siano P, Loia V, Tommasetti A, Catalão J. A review of smart cities based on the internet of things concept. Energies. 2017;**10**:421

[52] Baccarelli E, Naranjo P, Scarpiniti M, Shojafar M, Abawajy J. Fog of everything: Energy-efficient networked computing architectures, research challenges, and a case study. IEEE Access. 2017;**5**:9882-9910

[53] Ahad MA, Paiva S, Tripathi G, Feroz N. Enabling technologies and sustainable smart cities. Sustainable Cities and Society. 2020;**61**:102301. DOI: 10.1016/j.scs.2020.102301. ISSN 2210-6707

[54] Maksimovic M. The role of green internet of things (G-IoT) and big data in making cities smarter, safer and more sustainable. International Journal of Computing and Digital Systems. 2017;**6**: 175-184

[55] Amponsem J, Sänger N, Graf, MC. A youth perspective ongreen local urban futures. In City Preparedness for the Climate Crisis. Edward Elgar Publishing; 2021. pp. 331-343

[56] Cantador, I., Bellogin, A., Cores-Cediel, M. and Gil, O. Personalized recommendations in e-participation: Offline experiments for the Decide Madrid platform. In Proceedings Of The International Workshop On Recommender Systems For Citizens. pp. 1-6 (2017)

[57] Morabito V. Big data and analytics for government innovation. Big Data and Analytics. 2015:23-45

[58] Su, K., Li, J. and Fu, H. Smart city and the applications. In 2011 International Conference On Electronics, Communications And Control (ICECC). pp. 1028-1031 (2011)

[59] Hashem I, Anuar N, Gani A, Yaqoob I, Xia F, Khan S. MapReduce: Review and open challenges. Scientometrics. 2016;**109**:389-422

[60] Myeong, S., Jung, Y. & Lee, E. A study on determinant factors in smart city development: An analytic hierarchy process analysis. Sustainability. **10**, 2606 (2018,7,25). Available from: http://www.mdpi.com/2071-1050/10/8/2606

Chapter 5

The Effect of Smart City on the Promotion of Entrepreneurship

Jae Eun You and Jong Woo Choi

Abstract

As smart cities are early markets, there are many opportunities for new services, technologies, and platforms to enter. The smart city ecosystem is also necessary for sustainable smart cities. The smart city plan can serve as an opportunity for startups to take a new leap forward at a time when various regulations and fierce competition are entering a stagnant period. The brilliant ideas of prospective entrepreneurs have become new vitality in the smart city field. Prospective start-ups will come up with solutions and service ideas for a more convenient and safe life by analyzing market demand and gaps. If there were no various regulations in the smart city and various support policies, chances for start-ups would increase.

Keywords: smart city, entrepreneurship, start-ups, innovations, strategies

1. Introduction

From the beginning, smart cities received a lot of attention from both developed and developing countries regardless of their national technology level. Developed countries are promoting smart cities for optimal utilization and efficient management of existing urban infrastructure to solve urban problems, and developing countries have been interested in skipping technology levels by applying new urban infrastructure construction [1].

On the other hand, the general argument is that the concept of smart city varies depending on the economic level or policies of each country, region, or city, and there is no universally available concept. For example, the International Telecommunications Union [2] investigated the current status of global smart cities and identified 116 similar but different smart city definitions.

Recently, as smart city discussions continue, smart cities generally agree that information and communication technologies are applied to urban spaces to efficiently utilize urban resources [3].

In response to the fourth industrial revolution caused by the development of IT technology, the word smart city means a combination of urban space and technology [4, 5], transforming the city into a new technology creation space. Due to the hyperconnection of virtual space and physical space, an industry that maximizes efficiency by collecting people, assets, and data based on platforms and sharing goods and services, as in the case of Uber and Airbnb, is rapidly growing. In addition, the fourth industrial revolution is difficult to predict due to its rapid development and constant convergence with other fields [6], and regulatory reform for the acceptance and spread of fast-growing new technologies is an important role for the government and a place to experiment [6]. In response to the fourth industrial revolution, the smart city approach of foreign countries also deals with "data-based platform access and demonstration in urban space," and Korea is also introducing such smart city policies.

Smart cities have something in common in that they actively utilize the characteristics of making the most of the city's resources based on information and communication technology, but there are some differences in the background, major means, and goals. As can be seen from the individual appearance of smart cities, mobility minimization, expansion of civic exchanges, efficient solutions compared to investment, data and platforms, urban demonstration, and regulation are suggested as major means. Due to different backgrounds, the scope of smart cities is expanding, and as a result, difficulties exist in promoting smart city policies. In order for the smart city policy to be successfully promoted, it is necessary to clarify the goals and means to be achieved.

Although there are differences among works of literature, the components of smart cities are largely presented in the technical sector, the human resource sector, the institutional sector, and the innovation sector. Hwang [7] presented seven layers of smart city components such as urban infrastructure, ICT infrastructure, and spatial information infrastructure, focusing on the technology sector.

Smart cities can be largely divided into technology and infrastructure sectors, institutional sectors, and human resources sectors. The technology and infrastructure sectors include physical "urban infrastructure" such as roads and bridges, information and communication infrastructure such as communication networks, and information and communication-related technologies and platforms represented by ICBM. In the institutional sector, the areas covered by smart cities are wide, and due to the characteristics of convergence and complexity by field, it shows high importance at the stage of actual management and operation of smart cities. The human resource sector acts as an important means of creating innovative services and achieving a smart city ecosystem by promoting innovative ideas, cooperation with the private sector, and citizens' participation, reflecting the characteristics of rapidly developing and evolving information and communication technology.

The components of smart cities can be largely divided into infrastructure, data, services, and institutional sectors, and are included in seven detailed elements for each sector. The infrastructure sector includes urban infrastructure that can apply related technologies and services as physical and technical elements for building a smart city, ICT infrastructure that connects the entire city, and spatial information infrastructure necessary to fuse real space and cyberspace. The data sector is an area related to the production and sharing of data necessary to develop and operate new urban services, and IoT technology is a key element. The service sector is an area that provides actual urban services and includes algorithms for data utilization, reliable services, and urban innovation elements that are the basis for social and institutional.

2. Creative smart city composition

2.1 Open lab operations

Based on NB-IoT technology, this place provides a cooperative system and opportunity to develop innovative devices and creative services with mobile carriers in the IoT market. By securing compatibility between various smart city products and services, it aims to improve user convenience, promote new product and service development, and create a smart home industry ecosystem that coexists with small and medium-sized companies. It supports the development of open smart city linkage technology and services. It helps develop IoT interworking technologies for interoperation between various products and devices of various manufacturers.

In addition, IoT convergence service models are discovered through collaboration between large and small businesses based on interlocking, and open smart city testbeds are operated and demonstrated. It demonstrates the interoperability, compatibility, and commercialization appropriateness of various products and services. It also provides consulting services for commercialization of related small and medium-sized enterprises. It prepares and distributes standards for open smart city linkage technology. It supports the operation of open IoT consultative bodies that participate in industry, academia, and research.

Open lab actively supports development companies and manufacturers to commercialize ideas by supporting development spaces and 3D printers, including NB-IoT communication modules and development boards. Open lab receives business proposals for innovative devices and creative services to revitalize the IoT ecosystem.

2.2 Start-up incubating

It will build a “start-up incubation support platform” that can be operated continuously by combining hardware elements such as space, facilities, and equipment for cooperation with companies in industrial complexes with software elements such as research resources and capabilities.

Spigel [8] classified the ecosystem of start-ups into cultural, social, and physical dimensions, and saw it as a process in which components of three dimensions interact with each other and resonate [9]. The cultural dimension refers to a risk-taking entrepreneurial culture or a successful model to benchmark, the social dimension includes social networking within the region and mentors that companies must follow, and the physical dimension encompasses national or local governments, infrastructure, research, and investment institutions [8]. This means that policy should create an environment in which entrepreneurs can start their own businesses without fear of failure. Fails in the growth of companies and entrepreneurs need to not become a burden on the economy to help.

Classification	Contents
Supporting Target	Regional-specific convergence VR and AR technology-based enterprises and startups
	• Enterprises in specialized industries and content production companies with 5G, VR, and AR technologies
	• VR and AR content tasks composed of consortiums with local companies
Supporting field	Specialized industries such as aviation, IoT, and 5G-based VR and AR contents
	• 5G-based content: High-capacity, high-quality content types that enable uninterrupted real-time experience and interaction with cloud servers and 5G wireless networks.

Table 1.
Fostering realistic content.

It is to complete a circular structure in which start-up companies can lead to new investments again through growth and spread. Therefore, early start-up companies should not have difficulty raising funds and securing talent.

2.3 Fostering realistic content

This project will be carried out as a project to foster new industries that foster innovative growth bases for industries by establishing 5G-based VR/AR platforms in connection with strategic industries, which are future growth engines. The main task is to create and utilize a development environment in VR and AR production base centers. It supports equipment and systems for technical support such as production and demonstration of convergence VR and AR contents (**Table 1**).

2.4 Building a cluster for revitalizing the unmanned aircraft (drone) industry

This project is a cluster construction project that builds a drone-only flight test site, in the landfill, in the Seoul metropolitan area, supports drone-related companies, and creates a space for civic experience (education) to create a foundation for fostering new industries.

The project will establish a foundation for revitalizing the drone industry by providing a drone-only test site so that safety tests can be freely performed for a certain period of time at the request of operators (users) to support various flight test conditions such as night and altitude. It will build infrastructure facilities such as control towers, maintenance warehouses, offices, and take-off and landing sites.

The expected effect of this creates a foundation for establishing the drone industry promotion policy of the Ministry of Land, Infrastructure, and Transport, thereby providing conditions for fostering new industries. It creates a synergy effect of expanding the base of unmanned aerial vehicles, such as attracting related industries. Policy and technical support through the support of experts from the Ministry of Land, Infrastructure, and Transport and the Korea Aerospace Exploration Institute will be needed (**Table 2**).

2.5 Supporting innovation growth in robot industry

There is a need for bold fostering and support policies to enhance the competitiveness of the robot industry. It is necessary to establish a mid-to a long-term plan to foster the robot industry as a specialized industry by setting five major policy directions.

Stage	Content
1. Securing the site	• Establishment of drone cluster construction plan and securing site
	• Completion of a drone-only flight test site
2. Perform Design	• Perform basic design for drone cluster construction
	• Supporting drone flight test sites and creating a citizen experience space
3. Operation	• Corporate support for drone clusters and operation of citizen experience space

Table 2.
Building a cluster for revitalizing the unmanned aircraft (drone) industry.

Specifically, it will create a representative robot ecosystem through the successful creation of Robot Land. It is necessary to establish a support system to discover and intensively foster the specialized robot field and foster the robot industry by supporting the innovative growth of robot companies and establishing and spreading robot culture among citizens.

The main task is to create infrastructure to foster the robot industry such as Robot Land. By creating the Robot Land, robot industry promotion facilities will be activated, and robot test and certification support centers will be established. It fosters robot start-ups and supports the innovative growth of robot companies.

Furthermore, it spreads robot culture among citizens. Through this, cultural projects that can raise awareness of robots and induce citizens' interest are carried out. Robot characters and content are developed and citizen participation in robot competitions is held.

Also, specialization creates the demand for robots. Through supply–demand matching, conditions for creating and revitalizing the robot industry ecosystem are created. Improve the organizational system and system for upgrading the robot industry system. Establish a cooperative system for robot companies and strengthen system improvement and support organizations (**Table 3**).

2.6 Recruitment and operation of private start-up training institutions

It is necessary to select a private fostering institution with the capability and infrastructure to discover and foster big data and AI-based startups that will lead to new industries by linking smart cities and bio-fusion. It supports private organizations

Strategy	Content
Creating a robot ecosystem	Development of robot-related industries, businesses, theme parks, housing, and commercial facilities
	Building a cluster of robot industries by attracting and revitalizing robot industry promotion facilities (Robot Tower, Robot R&D Center)
	Based on standards and technical standards, testing, inspection, calibration, and certification services such as reliability, performance, and safety of Korean robot companies are provided
Innovative growth of robot companies	Robot start-up discovery, start-up space operation, start-up, and job support
	New R&D Support for Robot Companies
	Support for patenting, testing, and certification of robot products
	Support for participation in promising domestic and foreign exhibitions in the robot field
	Networking support for robot companies
	System Improvement and Policy Development for Robot Industry Development
Spreading robot culture to citizens	Development of robot characters, establishment of robot experience center, and holding robot-related competitions robot idea contest, promote robot-related cultural events
	Robot training, camp operation

Table 3.
Supporting innovation growth in robot industry.

and companies that have the capacity and infrastructure to support startups and meet the qualifications.

As for the contents of the support, space to discover and foster startups and common and specialized startup programs (evidence, R&D, education, etc.) are supported. The selection of tenant companies shall be promoted through the formed committee. Rent for space will be provided free of charge for up to 5 years to tenant companies, and major fostering industries through common and specialized startup programs will be selected as fourth industry-specialized fields, such as big data, AI, blockchain, and IoT.

3. Smart city strategy

Currently, many local governments are promoting or planning to promote platform-oriented smart cities. It is also aware of the importance of the platform when promoting smart cities, but it is still in the early stages of settling down where the results of the smart city platform are not as expected.

The distribution of urban integrated operation centers, which can be said to be the physical infrastructure of platform-oriented smart cities, is held by many local governments from the beginning, and local governments using self-communication networks are also showing a high rate. In addition, an integrated platform with a public nature is also expanding its construction through distribution and diffusion projects to local governments every year. While the distribution of physical infrastructure and integrated platforms is being structured or rapidly spreading, the data linkage and utilization sector, which is the core of platform-oriented smart cities, is still insufficient, causing low achievement of platform-oriented smart cities. In order for platform-oriented smart cities to be implemented, the most important direction is to promote policies so that more data can be linked and utilized, and in addition, platform policy directions should be established so that various subjects can participate and spread spatially. In fact, it will be possible to achieve the goals pursued by platform-oriented smart cities only when public and private data are linked and integrated, and at the same time, their joint use gradually spreads locally.

3.1 Seeking ways to link with private data

The current level of domestic smart cities is limited to linking and utilizing public data between different service fields. In the case of overseas, there are some cases in which private big data is used together with public data based on the platform of the urban integrated operation center. The biggest problem in linking private data with the platform in the domestic urban integration operation center is linked to the Personal Information Protection Act. Local governments' opinions also argue that the integrated platform should be partially opened to increase its utilization.

Due to the existence of personal information in the integrated operation center and platform, local government officials are reluctant to actively develop and link solutions, and they are burdened by the law in terms of introducing and linking smart city solutions. While interest in platform-linked smart cities is mentioned, as the highest target type of local governments, the fiscal investment sector is the lowest so far. Therefore, there is a need to promote platform linkage as a national project so that smart city solutions suitable for the characteristics of local governments can be discovered. If a pilot project linking private and public data is promoted by combining various newly introduced regulatory sandboxes, local governments will be able to push for a more distinctive smart city with the central government's budget support.

Among the current government's smart city promotion strategies, problem-solving smart city-type urban regeneration targeting areas with innovative industries through national pilot cities have clear goals. However, in the case of existing urban areas, including new cities, where construction has been completed, there is an area where the identity of the goal is ambiguous. Therefore, as one of the goals of upgrading smart cities, it is necessary to seek ways to upgrade to data-oriented smart cities in the form of discovering and distributing private and public data-linked services as well as the current integrated platform distribution project.

In terms of technology, the current closed smart city platform, which is disconnected from the external network, should be upgraded to an open smart city platform so that information can be disclosed and provided. Due to regulations related to personal information protection, the current smart city platform is cut off from the outside world. Since there are definitely limitations in linking only public services, there is also a need to convert to a platform form in which data can be opened in order to expand its functions. Since it is practically impossible to switch from a closed platform to an open platform, due to current regulations, the most realistic alternative would be to apply it first using the regulatory special cases of the national pilot city currently being promoted and gradually spreading after a successful demonstration.

3.2 Smart city with innovative space creation

Smart cities that create innovative spaces can be seen as a type that lags behind in Korea compared to advanced countries abroad. The creation of innovative spaces includes demonstration and inter-city networks, and in the case of demonstration, there are many factors to consider, such as deregulation, goal-oriented performance indicators, open data policies, living labs based on civic participation, and new industries. Since inter-city networks simultaneously play a role as knowledge exchanges and potential overseas markets, global network support is needed at the national level.

Since discussions on innovative space creation-type smart cities have recently begun in the domestic smart city policy sector, it is necessary to consider various strategies. As can be seen from the results of the local government's survey, there are few local governments that are currently promoting innovation space creation in smart cities in Korea, but as can be seen from domestic keyword analysis, keywords related to innovation have increased significantly over the past 2 to 3 years.

3.3 Introduction of practical regulatory sandboxes

With the recent revision of the Smart City Act, national pilot cities have introduced and operated regulatory sandboxes. However, domestic regulatory sandboxes have the nature of regulatory exceptions for prescribed industries such as self-driving cars and drones, which the government believes have potential rather than discovering unexpected new innovative industries. Regulatory sandboxes promoted in Japan are more comprehensive than national pilot cities in Korea because they try to introduce a method of granting regulatory deferral by deliberating on proposals from private companies that want to test new industries. Therefore, it is necessary to expand the subject of regulatory grace in the direction of Japanese-style regulatory sandboxes that can further expand the scope of regulatory sandboxes.

As in the case of Japan, in order to implement regulatory sandboxes, the legal system needs to be improved in the short term, and it is also necessary to consider the management and operation of regulatory sandboxes. First, an organizational system related to the regulatory sandbox should be formed after the revision of the law in the form of a complete regulatory sandbox in the current form of special regulations. Two support organizations are needed to manage and operate regulatory sandboxes.

First, it is necessary to evaluate private companies' proposals for testing new industries in the regulatory sandbox, and at the same time, to monitor the performance and side effects of the new industry testing process of private companies. The role of the support center is to establish standards for allowing proposals, review the contents of proposal evaluation, determine acceptance levels, continuous monitoring of performance and side effects in the experiment process, and secure and promote markets for new industries that have been successfully demonstrated.

Second, a one-stop regulatory improvement center is needed. Most of the new industries that are newly tested in the regulatory sandbox are industries that are difficult to spread to other regions due to regulations. Therefore, it is necessary to operate a regulatory improvement center that monitors the tested process and supports the promotion of regulatory improvement at the same time if there is a high possibility of success. In general, it takes about 2 years until regulations are improved and legalized, so tests and regulatory improvement preparations need to be carried out in parallel so that they can be used immediately after a successful demonstration [10].

3.4 Promotion of special industry classification in smart urban industries

On the other hand, although the concept of smart cities is ambiguous and the definition and classification of related industries are unclear and extensive, efforts need to narrow the scope of smart city industries to revitalize the smart city innovation ecosystem. As seen by integrated platform construction companies, the standard industry classification code is a mixture of manufacturing and software developers, and there are limitations that it is difficult to define the smart city industry based on the existing classification criteria. Therefore, it is necessary to quickly establish special industry classifications related to smart cities by referring to special classifications, such as the spatial information industry, and gradually supplement them in the future.

3.5 Improvement of smart city certification and standardization system

It is expected that new services will be derived by converging and combining existing services in relation to the smart city industry. Standardization should be supported, so that different services are linked and compatible under these conditions of convergence. Therefore, when various smart city services are developed, it is necessary to present guidelines related to standard procedures and frameworks to enable convergence, complexity, and interoperability based on compatibility [11]. These efforts are related to entering not only domestic but also overseas smart city businesses, and for this, international cooperation related to smart city standardization is also needed.

3.6 Social and cultural innovation

Sociocultural innovation can be said to be the most abstract and difficult to present measures compared to the innovations of the assets discussed above. In the case of networking assets discussed in urban innovation spaces, a culture that allows people to interact and talk with each other may be a more appropriate model for the West than for the Asian region. However, in Korea, if various community activities are activated and an environment where people can gather is created, the possibility of physical networking can increase. In fact, the prerequisite in terms of revitalizing platforms and networking assets is that cities should play a role as a place where various people meet, share opinions, and communicate offline as well as on online platforms.

3.6.1 Fostering an adventurous entrepreneurial spirit

First of all, socio-cultural innovation is needed in relation to the training of talents who are the basis of networking. It was found that domestic companies lack entrepreneurship to enjoy the adventure without fear of risk, and job seekers prefer stable jobs (Asan [12]). It can be seen that this trend of job preference does not solve the most fundamental problem of forming a smart urban innovation ecosystem. Therefore, it is necessary for the public and private sectors to cooperate to educate entrepreneurship so that the start-up culture of enjoying and attempting new adventures can spread (Asan [12]).

In addition, it is related to socio-cultural changes, and even if necessary manpower is trained in many universities, their careers cannot be connected to related majors and there is a problem of preferring stable jobs such as public officials. As such an example, a survey of high school students' future hopes showed that 50% preferred stable jobs such as public officials and professionals (Asan [12]). For this reason, it is necessary to recognize the perception that business activities related to start-ups can be stable for those who are worried about their careers. This direction can be resolved in a way that the government bears the failure, as previously discussed in the policy related to the revitalization of start-ups.

3.6.2 Promoting citizens' awareness of participation in smart city policies

If there is a difference between the smart urban innovation ecosystem and the general economic innovation ecosystem, it may be related to improving the quality of life of citizens. In other words, in order to revitalize smart urban innovation ecosystem, solving urban problems recognized by citizens, and creating a better city is a condition for revitalizing the smart urban innovation ecosystem. Therefore, it is necessary to collect opinions from citizens and create a system so that they can participate in urban policies and contribute to the city by providing urban problem solutions.

The way to influence citizens' smart city policies is to use civil petition data. However, it is necessary to introduce a direct democracy method rather than an indirect method using civil petition data to create a system so that citizens can directly propose urban policies and actively engage in urban management, and encourage citizens to participate in policies. In terms of technology, digital democracy using blockchain technology is under discussion, and there are cases that implement this. If direct democracy such as digital democracy is expanded, citizens' interest in policy participation could increase.

4. Conclusions

This study presented a framework for establishing the concept and analyzing components of the smart urban innovation ecosystem in detail through literature research. The smart urban innovation ecosystem framework includes platforms,

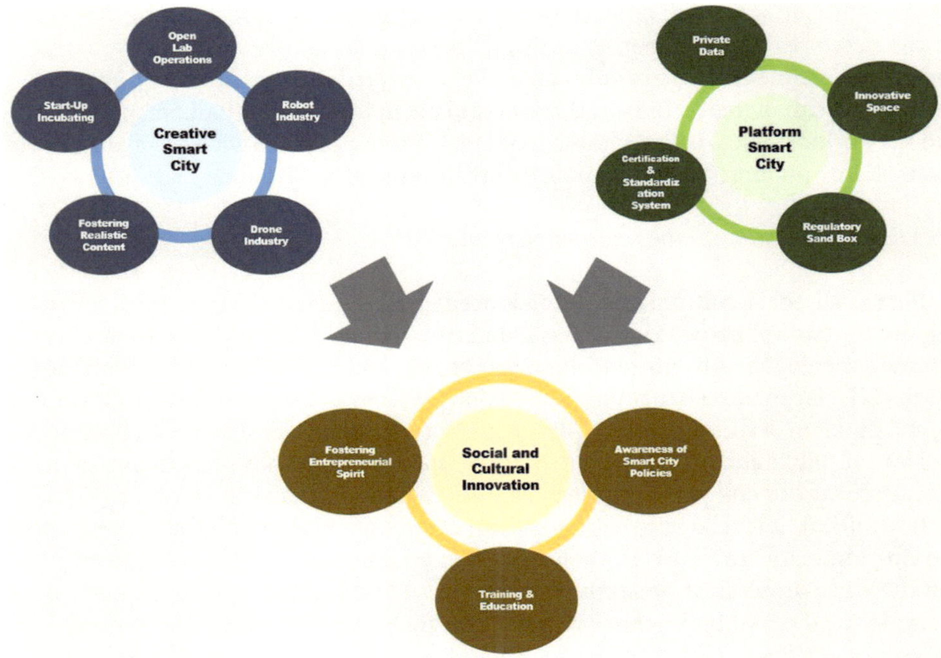

Figure 1.
The framework of the smart urban innovation ecosystem.

physical assets, virtual assets, human assets, economic assets, institutions, and social culture as components, among which platforms, that is, networking assets, are the most important (**Figure 1**). Here, the platform is a platform as a city mentioned in a smart city and refers to both a digital platform in a virtual space and a platform in a physical space and a virtual space that encompasses networking assets discussed in an urban innovation space.

Based on the smart urban innovation ecosystem framework, this study discussed ways to improve the innovation of each innovation ecosystem component and revitalize the overall innovation ecosystem, focusing on the platform. As the term ecosystem implies, which means the circulation of matter and energy, there is a connection between each innovation ecosystem component, and if the connection between them is not smooth, the spread of innovation and innovation of the components is difficult to occur. For example, human assets affect not only virtual assets, such as the collection of big data and analysis using artificial intelligence in terms of technology development, but also economic assets in terms of entrepreneurship and smart urban industry.

In the case of the startup ecosystem related to the innovation ecosystem in terms of economy, it may seem related to the industrial economy, but as seen in Urban Tech, it is an important element of the smart urban innovation ecosystem in that there are many startups related to urban problem-solving. These startups emerge, spread, and lead to innovation in physical and virtual assets. To lead this creativity, talent needs an entrepreneurial spirit to enjoy the adventure without fear of risk, which must be accompanied by socio-cultural innovation that supports new attempts and helps them recover even if they fail.

In terms of institutional aspects, existing smart city policies have carried out various institutional support projects such as infrastructure construction, R&D projects for technical support, and talent training projects. It is true that such institutional support contributed to the development of domestic smart cities, but the private sector and citizens were passive in the process of implementing these policies due to the government-led top-down ordering method. In the future, there should be institutional support so that the private sector and citizens can play a leading role in establishing smart city policies. In addition, it is necessary to improve various regulations, such as improving the Personal Information Protection Act, and ultimately seek ways to realize comprehensive negative regulations. In addition, in order to present specific measures for the creation and revitalization of the smart urban innovation ecosystem, it is necessary to preemptively present the concept and scope by introducing a special classification of the smart urban industry.

Ultimately, it is necessary to activate the platform so that the innovations between these assets can be smoothly connected. Until now, it can be said that discussions are underway rather than implementing smart cities as platforms. As an ideal platform, smart cities are not only virtual city platforms that collect, analyze, and simulate urban data, but also online and offline platforms that integrate physical and virtual environments by expanding physical networking to exchange ideas through offline community activities.

Author details

Jae Eun You[1] and Jong Woo Choi[2]*

1 Graduate School of Public Administration, Seoul National University, South Korea

2 Digital Agriculture and Technology Management Lab, Department of Agricultural Economics and Rural Development, College of Agriculture and Life Sciences, Seoul National University, South Korea

*Address all correspondence to: youchoi817@snu.ac.kr

References

[1] United Nations Economic and Social Council. Smart Cities and Infrastructure. Commission on Science and Technology for Development: Genova; 2016

[2] ITU-T. Smart sustainable cities: An analysis of definitions. ITU-T focus group on smart sustainable cities Technical Report. 2014

[3] Neirotti P, De Marco A, Cagliano AC, Mangano G, Scorrano F. Current trends in smart city initiatives: Some stylised facts. Cities. 2014;**38**:25-36

[4] Hollands R. Will the real smart city please stand up? Intelligent, progressive or entrepreneurial? City. 2008;**12**(3):303-320

[5] Komninos N. Intelligent Cities: Innovation, Knowledge Systems and Digital Spaces. London: Routledge; 2002

[6] Schwab K. The Fourth Industrial Revolution: What it Means, How to Respond. 2016

[7] Hwang JS. Smart city project national strategy by evolving trends and issues. Journal of the Korean Institute of Communication Sciences. 2017;**34**(8):14-18

[8] Spigel B. The relational organization of entrepreneurial ecosystems. Entrepreneurship Theory and Practice. 2017;**41**(1):49-72

[9] Lee JW, Kim SW, Kim YH, Lee YJ, Baek SI, Kwon KH, et al. 2018 Entrepreneurship and Startup Ecosystem Monitoring. Sejong: Science and Technology Policy Institute; 2019

[10] Zhou Y, Kankanhalli A. AI regulation for smart cities: Challenges and principles. In: Smart Cities and Smart Governance. Cham: Springer; 2021. pp. 101-118

[11] National Institute of Technology. R&D Roadmap Smart Based on 2019 New Growth Industry Standards City. Chungbuk Seoul: National Institute of Technology and Standards Korea Standards Association; 2019

[12] Asan Foundation. Startup Korea to Revitalize the Startup Ecosystem Seoul: Asan Foundation. Seoul: Korea Startup Forum; 2019

Chapter 6

Research on Technology Governance of IoT Smart City in Yilan, Taiwan: Taking Intelligent Disaster Prevention as an Example

Chang-Wei Chai, Yu-Heng Huang and Tseng-Wei Chao

Abstract

An advanced smart building platform should meet the needs of humanization and provide the best information and communication technology integration capabilities to enhance the digital upgrading and transformation of the construction industry. From the perspective of the technology governance of the smart city in Yilan County, this study proposes the direction of the sustainable development of intelligence in Lan-yang area and discusses the intelligent disaster prevention of the IoT smart city in Yilan, Taiwan. In the study, data related to technology governance were collected through literature review and validation of empirical fire drills, including the meaning of smart city, the development process of promoting smart city in Taiwan, and technology governance and smart city development. The content and analysis of empirical fire drills demonstrated the specific achievements of the smart city technology governance development in Yilan County, Taiwan.

Keywords: technology governance, intelligent city, Internet of Things, intelligent disaster prevention, fire protection

1. Introduction

Last year, IEEE held the 2021 IEEE International Conference on Electronic Communications, Internet of Things and Big Data (ICEIB 2021) international symposium in Yilan, Taiwan, and one of the papers was selected as the Best Paper Award. The paper provides forward-looking AIoT (The Artificial Intelligence of Things) Fairy technical capabilities through the Infairy IoT (Internet of Things) browser platform, integrates different communication protocols, realizes the interconnection and interoperability of IoT standards, solves the smart home fragmentation, and meets the needs of intelligent buildings to keep pace with the times.

The study site is located in Yilan County, northeastern Taiwan. The location is small and beautiful, which is conducive to the development of the pilot smart city market. Through the smart application of technology governance and industrial development, it will bring a new and better life experience, expand the energy of technology governance and the IoT industry, and introduce the smart city market.

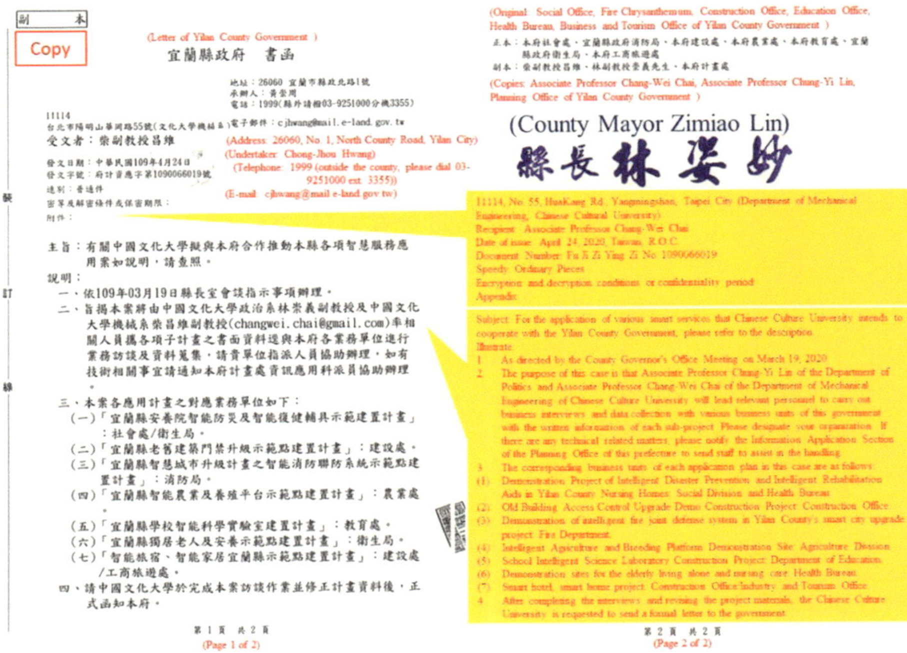

副本

Copy

(Letter of Yilan County Government)

宜蘭縣政府 書函

地址：26060 宜蘭市縣政北路1號
承辦人：黃崇周
電話：1999(縣外請撥03-9251000分機3355)
電子郵件：cjhwang@mail.e-land.gov.tw

(Address: 26060, No. 1, North County Road, Yilan City)
(Undertaker: Chong-Jhou Hwang)
(Telephone: 1999 (outside the county, please dial 03-9251000 ext. 3355))
(E-mail: cjhwang@mail.e-land.gov.tw)

11114
台北市陽明山華岡路55號(文化大學機械系)
受文者：柴副教授昌維

發文日期：中華民國109年4月24日
發文字號：府計資應字第1090066019號
速別：普通件
密等及解密條件或保密期限：
附件：

主旨：有關中國文化大學擬與本府合作推動本縣各項智慧服務應用案如說明，請查照。

說明：

一、依109年03月19日縣長室會談指示事項辦理。

二、旨揭本案將由中國文化大學政治系林崇義副教授及中國文化大學機械系柴昌維副教授(changwei.chai@gmail.com)率相關人員攜各項子計畫之書面資料逕與本府各業務單位進行業務訪談及資料蒐集，請貴單位指派人員協助辦理，如有技術相關事宜請通知本府計畫處資訊應用科派員協助辦理。

三、本案各應用計畫之對應業務單位如下：

(一)「宜蘭縣安養院智能防災及智能復健輔具示範建置計畫」：社會處/衛生局。

(二)「宜蘭縣老舊建築門禁升級示範點建置計畫」：建設處。

(三)「宜蘭縣智慧城市升級計畫之智能消防聯防系統示範點建置計畫」：消防局。

(四)「宜蘭縣智能農業及養殖平台示範點建置計畫」：農業處。

(五)「宜蘭縣學校智能科學實驗室建置計畫」：教育處。

(六)「宜蘭縣獨居老人及安養示範點建置計畫」：衛生局。

(七)「智能旅宿、智能家居宜蘭縣示範點建置計畫」：建設處/工商旅遊處。

四、請中國文化大學於完成本案訪談作業並修正計畫資料後，正式函知本府。

第1頁 共2頁
(Page 1 of 2)

(Original: Social Office, Fire Chrysanthemum, Construction Office, Education Office, Health Bureau, Business and Tourism Office of Yilan County Government)

正本：本府社會處、宜蘭縣政府消防局、本府建設處、本府農業處、本府教育處、宜蘭縣政府衛生局、本府工商旅遊處
副本：柴副教授昌維、林副教授崇義先生、本府計畫處

(Copies: Associate Professor Chang-Wei Chai, Associate Professor Chung-Yi Lin, Planning Office of Yilan County Government)

(County Mayor Zimiao Lin)

縣長林姿妙

11114, No. 55, HuaKang Rd., Yangmingshan, Taipei City (Department of Mechanical Engineering, Chinese Cultural University)
Recipient: Associate Professor Chang-Wei Chai
Date of issue: April 24, 2020, Taiwan, R.O.C.
Document Number: Fu Ji Zi Ying Zi No. 1090066019
Speedy: Ordinary Pieces
Encryption and decryption conditions or confidentiality period
Appendix

Subject: For the application of various smart services that Chinese Culture University intends to cooperate with the Yilan County Government, please refer to the description

Illustrate

1. As directed by the County Governor's Office Meeting on March 19, 2020
2. The purpose of this case is that Associate Professor Chung-Yi Lin of the Department of Politics and Associate Professor Chang-Wei Chai of the Department of Mechanical Engineering of Chinese Culture University will lead relevant personnel to carry out business interviews and data collection with various business units of this government with the written information of each sub-project. Please designate your organization. If there are any technical related matters, please notify the Information Application Section of the Planning Office of this prefecture to send staff to assist in the handling.
3. The corresponding business units of each application plan in this case are as follows:
   (1) Demonstration Project of Intelligent Disaster Prevention and Intelligent Rehabilitation Aids in Yilan County Nursing Homes: Social Division and Health Bureau
   (2) Old Building Access Control Upgrade Demo Construction Project: Construction Office
   (3) Demonstration of intelligent fire joint defense system in Yilan County's smart city upgrade project: Fire Department
   (4) Intelligent Agriculture and Breeding Platform Demonstration Site: Agriculture Division
   (5) School Intelligent Science Laboratory Construction Project: Department of Education
   (6) Demonstration sites for the elderly living alone and nursing care: Health Bureau
   (7) Smart hotel, smart home project: Construction Office/Industry and Tourism Office
4. After completing the interviews and revising the project materials, the Chinese Culture University is requested to send a formal letter to the government.

第2頁 共2頁
(Page 2 of 2)

Figure 1.
Smart city official documents approved by Yilan County Government.

This study has been approved by the local government of Yilan County with the approval letter number: 1090066019, as shown in **Figure 1**, smart city official documents approved by Yilan County Government. The approved research field goals include the following seven aspects: (1) Demonstration Project of Intelligent Disaster Prevention and Intelligent Rehabilitation Aids in Yilan County Nursing Homes: Social Division and Health Bureau, (2) Old Building Access Control Upgrade Demo Construction Project in Yilan County: Construction Office, (3) Demonstration of intelligent fire joint defense system in Yilan County's smart city upgrade project: Fire Department, (4) Intelligent Agriculture and Breeding Platform Demonstration Site in Yilan County: Agriculture Division, (5) School Intelligent Science Laboratory Construction Project: Department of Education in Yilan County, (6) Demonstration sites for the elderly living alone and nursing care in Yilan County: Health Bureau, (7) Smart hotel, smart home project in Yilan County: Construction Office/Industry and Tourism Office.

This chapter focuses on the third field "Demonstration of intelligent fire joint defense system in Yilan County's smart city upgrade project" and combines the IoT technology as the core technology to build a smart city in Yilan, especially the intelligent fire detection and alarm system in Yilan County.

2. Literature review

The concept of smart city originates from the concept of "Smarter Planet" proposed by IBM. A smart city is a measure for the transformation of a smart planet from concept to concrete implementation. Smart city construction is to develop and apply massive data, cloud computing, IoT, mobile information and communication technologies on the basis of intelligence, develop new urban concepts and governance methods, and improve urban operation efficiency. The construction of information and communication technology and environmental facilities will create a sustainable ecological environment, allow citizens and enterprises to enjoy a more comfortable and convenient environment, and enhance the competitiveness of the city. A smart city is defined as a city that performs well in terms of future economy, mobility, environment, citizenship, quality of life, and government [1].

Six thematic areas must be present and addressed in any Smart City proposal. They are: Smart Economy, Smart People, Smart Mobility, Smart Living, Smart Government, and Smart Environment, proposed by Boyd Cohen in the model, known as "The Cohen Wheel" [2]. Each subject area contains a set of city indicators.

Cities face tight budget, which have led to budget cuts in addition to cost-cutting measures. Therefore, smarter and resilient infrastructure is essential to oversee urban challenges and transform urban environments [3]. The integration of smart city technologies can help achieve the goal of smarter and resilient infrastructure [4, 5]. Currently, proposals for cost-effective solutions to data-based decision-making are best involving IoT-based technologies [6]. The IoT is one of the technological paradigms destined to exponentially increase the connectivity of various devices. The main advantage of IoT is its high impact on the daily behavior of potential users [7].

The IoT and Artificial Intelligence (AI) are current hot research topics due to recent industry achievements, and they have been shown to achieve better results in many disciplines such as automated factories, public surveillance, asset monitoring, waste management, weather monitoring, etc. Combining the IoT and AI is an effective way to intelligently upgrade existing information systems [8].

Infairy Technology Company has been engaged in the research and development of the IoT technology for more than 10 years. It integrates all communication standards and provides a smart IoT platform for true connectivity. The traditional IoT is limited by the problem of non-interoperability of standards, while Infairy Technology has not changed its hardware standards. Compatible with the original standards, coupled with the "standard interoperability" technology, solves the problems of the traditional IoT [9].

In Taiwan, there are many master's and doctoral dissertations related to smart city research. These relevant dissertations can be used as a reference for the literature review of this study [10–16].

In terms of promoting smart cities in Taiwan, the Economic Development Council of the Executive Yuan (now the National Development and Reform Commission) passed the "Third National Construction Design in the New Century (2009~2012)" in 2008, in which the main axis of national development policy of spatial reconstruction is the fifth project, namely "Smart Taiwan." Since 2014, the National Development and Development Council has actively promoted the overall development plan of smart land and proposed the overall development plan and planning strategy of smart land. The content and analysis of the development of science and technology governance in Yilan County and the empirical fire drills of smart cities show the specific achievements of technology governance and smart city development in Yilan County, Taiwan, as follows: [17]

1. Governments at all levels in Taiwan are actively promoting the research planning and pilot tests of smart land. For the development of Information and Communication Technology (ICT), the Yilan County Government provides a process

that takes into account the different aspects and characteristics of the problem, as well as various impacts and challenges, to form views and ideas that belong to Yilan. In the process of thinking about developing a smart country.

2. List the 104-year planning goals: In view of the problems faced by the urban development of Yilan County, through the operation of smart Yilan software and hardware, observe and understand the information environment of Yilan County's future development problems and propose solution to introduce big data and GIS-related technologies for planning "smart cities." Through big data, open data, GIS, and other processing and cloud computing, it provides intelligent information that can be applied to smart cities, whether it is urban governance, improving people's quality life experience, promoting the development of local industries, and extensive public services as the foundation of smart land and city governance.

The graphic text description of Smart Land Map of Taiwan (**Figure 2**) is as follows [17]:

1. Smart land = smart city + smart urban and rural areas (smart townships) + environmentally sensitive areas (mountain and sea monitoring)

2. The category of smart land: smart environment, smart economy, smart life, smart planning and governance, and smart society.

3. Smart country is an extension of smart city. ICT is introduced to assist in the development of cities and towns, as well as monitoring of mountains and coasts in environmentally sensitive areas. Advanced technology is used to improve the efficiency of government governance and the effective use of resources.

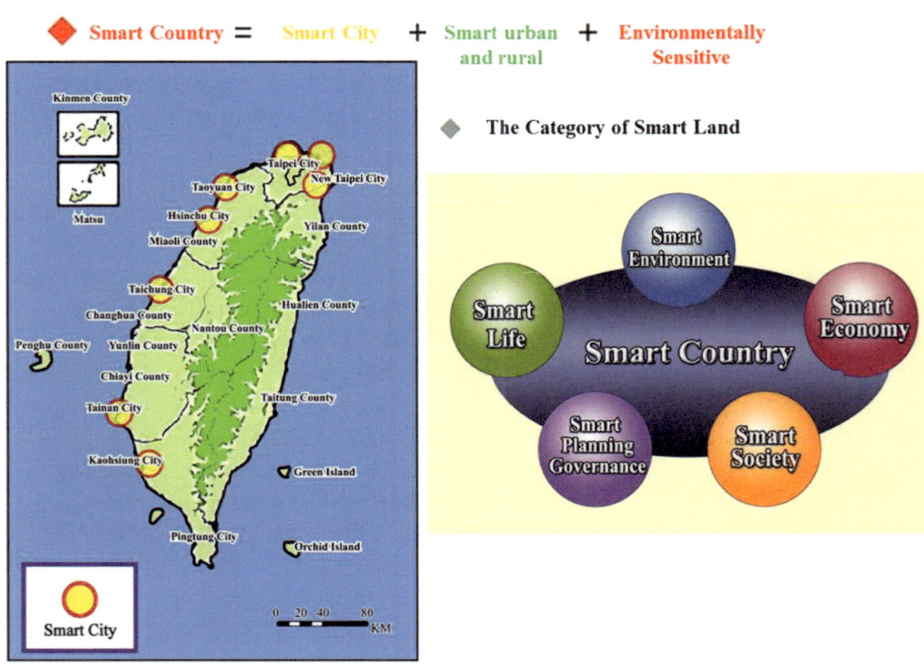

Figure 2.
Smart Land Map Yilan County in Taiwan [17].

DOI: http://dx.doi.org/10.5772/intechopen.106819

3. Research methodology

Knowledge production within the field of business research is accelerating at an alarming rate, while still being decentralized and interdisciplinary. This makes it difficult to keep up with the latest technologies and to be at the cutting edge of research, as well as to assess the collective evidence for a particular area of business research. This is why literature reviews are more relevant than ever as a research method. Traditional literature reviews often lack thoroughness and rigor and are conducted ad hoc, rather than following a specific methodology. Therefore, questions can be raised about the quality and credibility of these types of reviews. Literature review as a method of conducting research provides an overview of different types of reviews and some guidelines to how to conduct and evaluate literature review papers [18].

Empirical research method is a special form of scientific practice research. According to the needs of existing scientific theory and practice, propose a design, use scientific instruments and equipment, and determine conditions, through purposeful and step-by-step operations under natural conditions, according to the changes in phenomena accompanied by observation, recording, and measurement in the activity of causality between phenomena. The main purpose is to illustrate the relationship between various independent variables and a dependent variable.

Software engineering is more than technical solutions. It also largely deals with organizational issues, project management, and human behavior. For a discipline like software engineering, empirical methods are crucial because they allow the incorporation of human behavior into the research approach taken. Empirical methods are common practice in many other disciplines. One motivation for using empirical methods in software engineering research is that it is needed from an engineering perspective to allow for informed and well-grounded decision. Empirical research methods in software engineering continue with a brief introduction to four research methods: controlled experiments, case studies, surveys and postmortem analyses. These methods are then put into an improved context. The four methods are presented with the objective to introduce the reader to these methods so that they can select the most appropriate method in a specific situation [19].

This chapter applies research methods such as literature research method and empirical research method to analyze and discuss actual research cases, so as to fully present the technical governance application of smart city IoT intelligent disaster prevention in Yilan County, Taiwan. The research design is as follows:

3.1 Current status of disaster prevention

For the construction of the intelligent fire joint defense system in the study, after the firefighters entered the fire scene, because they could not grasp the internal space conditions, and the surrounding smoke was dense, the visibility was insufficient. , Lost direction, no way to get out of the predicament of export research, analysis, and control. In previous cases, such as the fire at the Jingpeng factory in Taoyuan City, Taiwan, the regrettable death of firefighters has occurred continuously. The reason why firefighters are so helpless is because they have too little information on disaster relief. In the past few decades, firefighters have only relied on heavy and heavy paper to grab pictures, which are not only difficult to preserve, but also inconvenient to carry. They failed to provide interpretation to front-line personnel in a timely manner, making firefighters almost blindly rush into the fire scene. In addition, the KTV fire in Taipei City, Taiwan, further highlights the importance of self-management of fire safety by the industry. Due to the complete shutdown of fire safety equipment such as smoke exhaust, sprinkler, alarm, and broadcasting systems, consumers cannot escape in time, causing heavy casualties. We should keep pace with the times, and only by combining technology can we improve the efficiency of disaster prevention and relief.

3.2 Intelligent disaster prevention solution strategy

In this research, an intelligent fire joint defense system is constructed in combination with the IoT technology, which provides real-time information for firefighters, so that firefighters can grasp the internal space situation after entering the fire scene. Using IoT technology to provide fire-related information for firefighters to interpret, reducing the casualties of firefighters due to insufficient information. In addition, actions that endanger public safety, such as the fire alarm system being closed privately by the operator, can also be communicated to the fire authorities through the real-time warning notification system to avoid the complete shutdown of fire safety equipment such as smoke exhaust, sprinkler, alarm, and broadcasting systems, resulting in consumption. The person cannot escape in time, resulting in the risk of serious casualties.

4. Research results

This research carried out a fire drill on June 20, 2020 at the Natural Beauty Dormitory, Wujie Township, Yilan County, Taiwan. The results of the relevant application of the IoT research were displayed in the fire drill. It is the first IoT technology disaster prevention exercise in the country. The fire chief of the county and city, the director of the Disaster Prevention Office of the Executive Yuan, and major media were all present. The research and development system combined industry-government-academic resources, led by the Fire Bureau of Yilan County Government and Chairman of Yilan County Council Jian-rong Zhang (Chief of Yilan County Volunteer Fire). The parliamentary secretary, Yu-heng Huang, was assigned to coordinate the liaison, participate in joint research and development, and introduce the first IoT-based intelligent fire protection system in Taiwan to improve disaster prevention and safety in Yilan County and more effectively reduce the danger of firefighters in the fire scene. The results of the relevant fire drills are as follows (**Figures 3–5**):

This exercise applies intelligent cloud platform of Infairy IoT browser technology as the IoT application infrastructure, combined with the firefighter's mobile positioning device of search action technology, intelligent smoke detection of Horing Lih industrial co., Ltd., temperature, gas wireless sensors, and other equipment to build a set of a security disaster prevention and relief system based on the IoT, research team members develop intelligent sensing technology, external sensors detect the operation of the trusted switchboard, and actively report abnormal behavior (ex. power is turned off) to create a smart fire city system. The system can actively report the operation of the fire safety equipment and the location of the fire in the event of a fire. It can receive the positioning signal in real time through the combination of the alarm device and the positioning device and send the plan of the fire site and the position of the firefighters back to the on-site commander's tablet computer. In order to avoid disasters in public places such as KTV in Taipei City, at the same time, this study cooperated with Taiwan Connection Co., Ltd., to enhance the LINE

DOI: http://dx.doi.org/10.5772/intechopen.106819

Figure 3.
Real photos of fire drills in Yilan County, Taiwan.

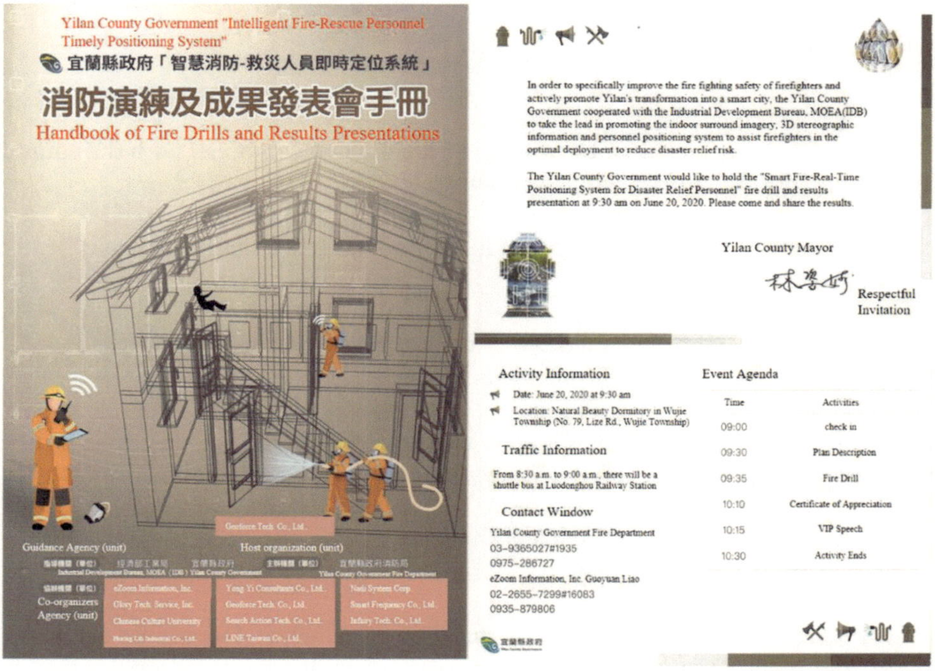

Figure 4.
Fire drill results presentation manual in Yilan County, Taiwan.

community function, and played the role of a new "Community Watch and Help." Through the LINE group notification function, the traditional "Community" Watch and Help Team has been advanced into a "Smart Community" Watch and Help Team Group, we can apply the IoT technology to build a smart city that makes people feel safe and secure at home.

Figure 5.
Invitation letter for the presentation of fire drill results in Yilan County, Taiwan.

The notification process for the use of IoT technology combined with the community software notification system is as follows:

1. When the smoke alarm is triggered in the home, LINE's "neighbor group" can be notified immediately.

2. LINE's "family group" can receive a notification immediately when the door is opened or when the elderly enters and exits.

3. When the electricity consumption is abnormal, the LINE "friend group" can receive abnormal electricity consumption information.

4. If a monitor is installed, watch it at the same time to transmit the monitor screen to each group synchronously.

The theme of this study exercise is "safety" and " relieved." During the exercise, through four major themes, this study will create an environment that will make Yilan County the most livable environment in Taiwan:

4.1 Intelligent disaster prevention linkage

When the smart sensor is triggered, the protective actions of firefighting through the Infairy smart cloud platform are as follows:

Open the windows of the home to exhaust smoke to avoid choking injuries and affect the escape time at the same time.

DOI: http://dx.doi.org/10.5772/intechopen.106819

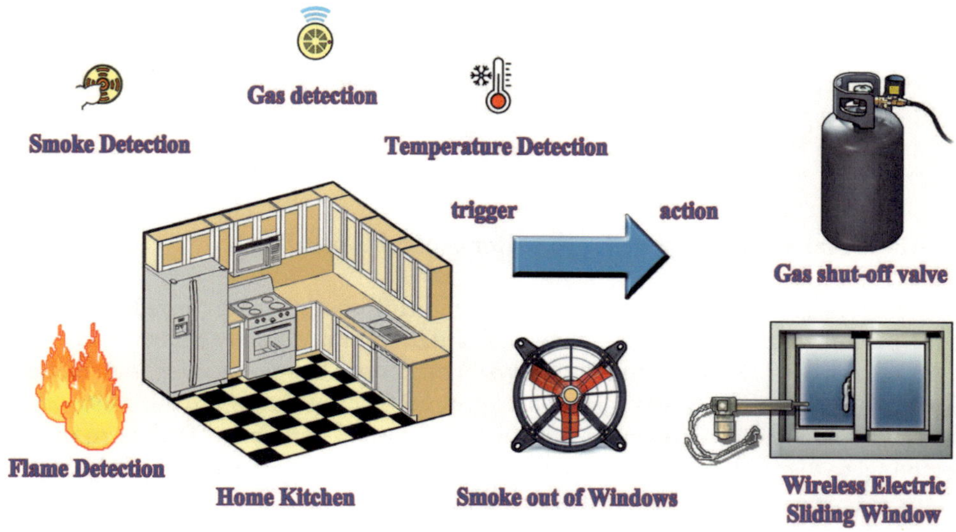

Figure 6.
Schematic diagram of intelligent disaster prevention linkage.

Turn off the gas to reduce the risk of increasing the fire due to gas leakage (**Video 1**, https://youtu.be/u9NxLUxXB9Q) (**Figure 6**).

4.2 Firefighters fire location

The personal safety of firefighters in the fire scene is also very important in this research. During the exercise, a fire location environment was built, and the

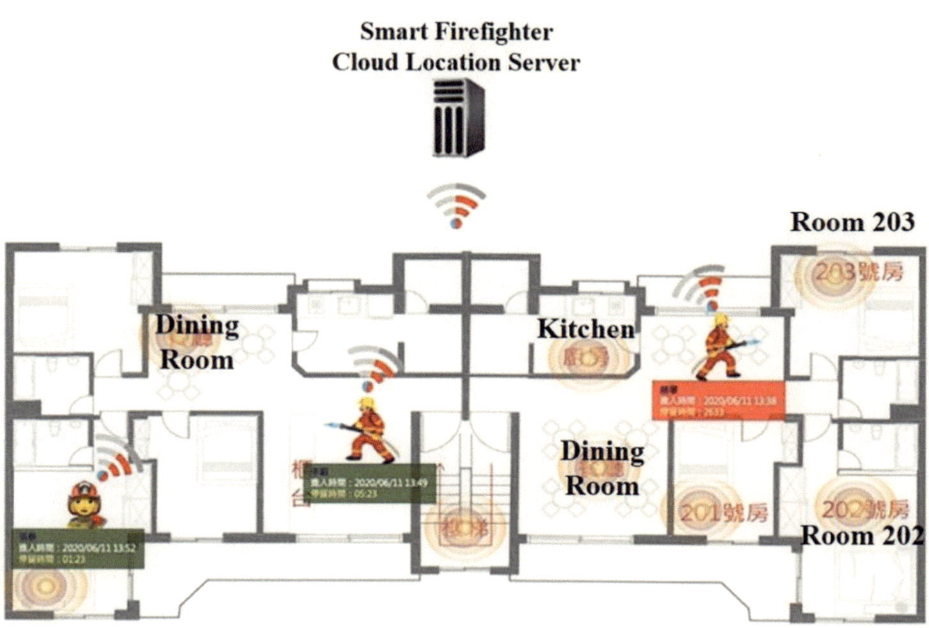

Figure 7.
Schematic diagram of firefighters' fire location.

indoor positioning App built into the firefighter's mobile phone automatically transmits the current location information to the command center. The real-time position, stay time, and travel trajectory of firefighters in the fire scene are easy to grasp and dispatch disaster relief, and at the same time, it can effectively reduce the danger of firefighters in the fire scene (**Figure** 7) (**Video** 2, https://youtu.be/U2KZqa-OCbE).

4.3 Abnormal notification of fire detector central control

When the trusted switchboard of the firefighting equipment is turned off, the detector immediately detects the abnormal situation, immediately transmits the message through the LINE communication system through the Infairy intelligent cloud platform, and notifies the designated background (such as the command center of the Fire station and related personnel), in order to immediately notify the responsible personnel to dispatch personnel to investigate and report to prevent the occurrence of disasters (**Figure** 8) (**Video** 3, https://youtu.be/wcB_iEEGjEM).

4.4 Community watch and help announcement

The LINE group of the social software is not only a channel for communication between modern people, but also can play a new role of "community watch and help." The relevant watch and mutual aid delivery procedures are as follows:

A. When the smoke alarm in the home is triggered, LINE's "Neighbor Group" can be notified immediately.

B. People can receive a notification immediately in "family group" of LINE when the door is opened or when the elderly enters and exits.

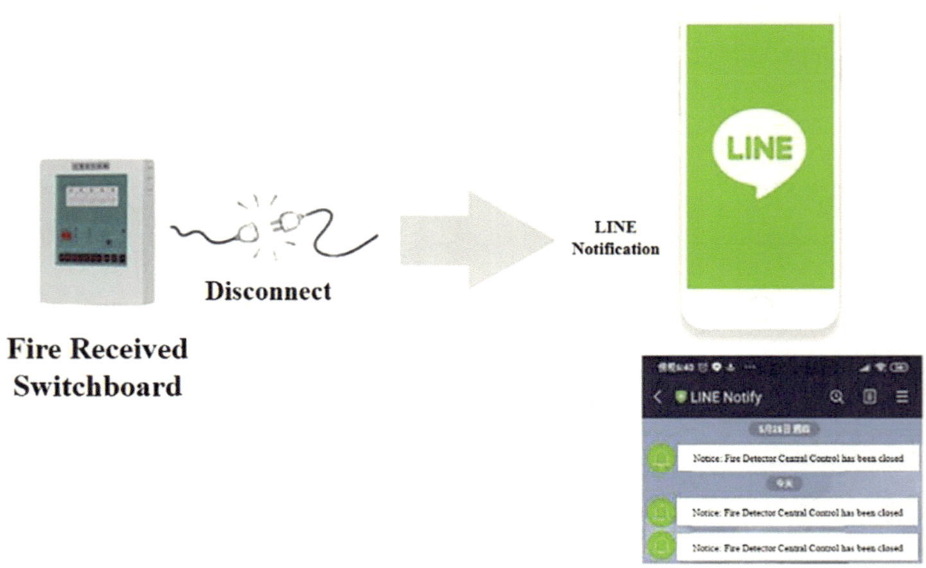

Figure 8.
Schematic diagram of abnormal notification of fire detector central control.

Figure 9.
Schematic diagram of community watch and help notification in LINE.

C. When the electricity consumption is abnormal, the "friend group" of LINE can receive abnormal electricity consumption information.

D. If a monitor is installed, LINE can simultaneously transmit the screen of the monitor to each group at the same time.

Through the LINE group notification function, the traditional "community" watch and support team has been advanced into a "smart community" watch and support group, and the use of IoT technology to build a smart city that makes people feel safe and secure at home (**Figure 9**) (**Video 4**, https://youtu.be/VAhDuhAUDVs)

The actual case of the application of IoT technology in technology governance successfully demonstrated in this research exercise. At the same time as the results of the exercise were displayed, Yilan County Mayor Zi-miao Lin led the announcement that Yilan County fully introduced IoT technology in agriculture, aquaculture, and the elderly living alone. Care, security update of old apartment access control, smart accommodation and other scenarios, and implement the IoT concept into the basic education courses of middle and primary schools, and establish seven major fields such as the IoT teaching practice laboratory in middle and primary schools, so that the IoT concept can be realized. Taking root downward, and taking Yilan County as a demonstration site, Yilan County will become the first smart city in Taiwan that can be felt by the public (**Figure 10**).

The research team has applied distance teaching technology to guide primary and secondary school students of remote indigenous tribes to learn new knowledge of science and technology, and the results have been unanimously affirmed and favored by teachers and students of remote indigenous tribes [20].

After entering the fire scene, the firefighters were unable to grasp the internal space conditions, the surrounding smoke was dense, and the visibility was insufficient. Although they were only a few steps away, they were disoriented due to the bad environment of the fire scene and could not get to the exit. There is too little disaster relief information. Over the past few decades, relying only on heavy paper to grab the pictures is not only difficult to store, but also inconvenient to carry.

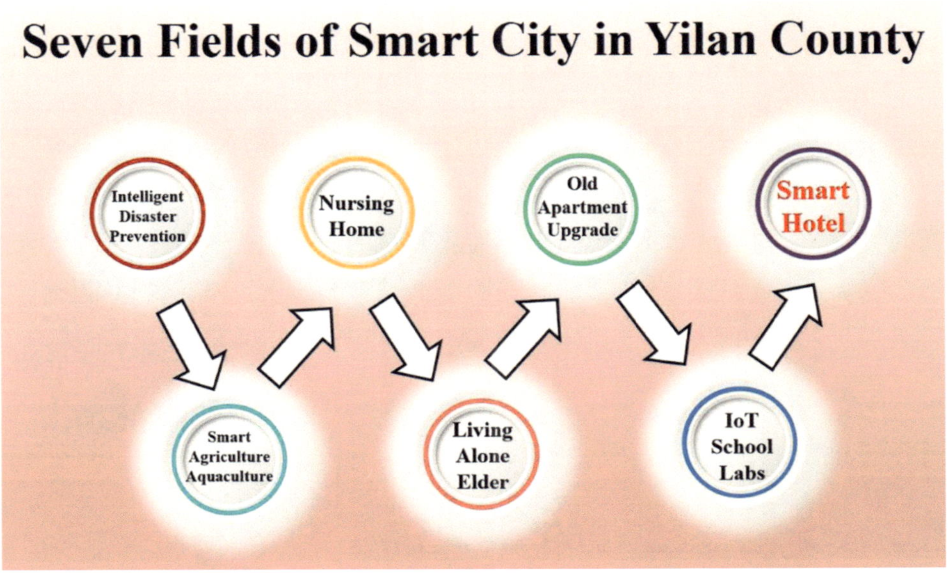

Figure 10.
Schematic diagram of the application of IoT technology in seven fields of smart city in Yilan County, Taiwan.

The tragedy cannot be repeated again. The Fire Department of Yilan County Government decided to cooperate with the team of Yizhong Information Co., Ltd., through the project of the Industrial Bureau of the Ministry of Economic Affairs to develop the "Smart Fire 3D Reality Control System," taking the lead in the country and establishing the basis for visualization. 3D map data, personnel entry and exit control, and indoor positioning system will present information such as the internal configuration of the building, firefighting equipment, and storage of dangerous goods in a three-dimensional manner, assisting in real-time grasp of disaster relief and effective deployment and human life search and rescue. Among them, the indoor positioning system can provide a warning function, reminding the commander to send personnel to follow the line for rescue and assist each firefighter to walk out of the fire safely. It has been gradually extended to 239 places such as hotels and restaurants, large exhibition venues, shopping malls, and dangerous goods factories within the jurisdiction, which is expected to greatly reduce the risk of disaster relief.

5. Conclusion

The introduction of IoT technology into smart cities will bring more convenience and fun to the city's daily work and life. In the future, urban intelligence will not be just cold steel, cement, and equipment, but will become a good partner in life, providing a safer, more convenient and energy-saving lifestyle. For example, the results of this research help fire and disaster relief to accurately grasp key information and effectively carry out disaster relief deployment and human life search and rescue operations. Among them, the indoor positioning system can provide a warning function, reminding the commander to dispatch personnel immediately, follow the line to rescue, and assist each firefighter to walk out of the fire safely. Relevant technologies can also be extended to hotels, restaurants, large exhibition venues, shopping malls, and dangerous goods factories in the county to reduce the risk of disaster relief for firefighters during disaster relief.

In addition, the establishment of a community watch and help LINE group in the research results can not only serve as a channel for communication between modern people, but also play a new role of "smart community watch and help," up to 1. When the smoke alarm in the home is triggered, "neighbor group" of LINE can be notified immediately. 2. "family group" of LINE can receive a notification immediately when the door is opened or when the elderly enters and exits. 3. When the electricity usage is abnormal, the "friend group" of LINE can receive abnormal electricity usage information. 4. If a monitor is installed, watch it at the same time to transmit the monitor screen to each group synchronously. By combining the IoT technology with the function of LINE group notification, the traditional "community" support team has been advanced into a "smart community" support group. The results show that people's homes are safer and more secure.

Acknowledgements

This study was supported by the Research Support Scheme of the Ministry of Science and Technology in Taiwan, R.O.C., grant no. MOST 110-2622-E-034-001-.

Conflict of interest

The authors declare no conflict of interest.

Notes/thanks/other declarations

The authors extend their deepest gratitude to the technical contributions of Chairman Dung-Ming Wu, former General Manager Zhi-Yuan Zhang, and President Ying Fang of Infairy Technology Co., Ltd., and the full assistance and cooperation of Yilan County Government. In addition, the authors are also very grateful to all those who participated in this research.

Author details

Chang-Wei Chai[1*], Yu-Heng Huang[1] and Tseng-Wei Chao[2]

1 Chinese Culture University, Taipei City, Taiwan

2 Shih Hsin University, Taipei City, Taiwan

*Address all correspondence to: changwei.chai@gmail.com

References

[1] Peek GJ, Troxler P. City in transition: Urban open innovation environments as a radical innovation. In: Proceedings of the 19th International Conference on Urban Planning, Regional Development and Information Society (REAL CORP). Vienna, Austria; 2014. pp. 1-10

[2] Ceballos GR. La viabilidad de un proyecto de Smart City como estrategia mercadológica. Caso: CUCEA. Red Int. Investig. Compet. 2018;**10**:629-646

[3] Novotný R, Kuchta R, Kadlec J. Smart city concept, applications and services. Journal of Telecommunication System and Management. 2014;**3**:117

[4] Smith SF. Smart infrastructure for future urban mobility. AI Magazine. 2020;**41**:5-18

[5] Ramamoorthy S, Kowsigan M, Balasubramanie P, Paul PJ. Smart city infrastructure management system using IoT. In: Role of Edge Analytics in Sustainable Smart City Development. Hoboken, NJ, USA: Wiley; 2020. pp. 127-138

[6] Dave E. Internet of Things, How the Next Evolution of the Internet Is Changing Everything. Cisco Internet Business Solutions Group (IBSG). 2011. Available from: https://www.cisco.com/c/dam/en_us/about/ac79/docs/innov/IoT_IBSG_0411FINAL.pdf [Accessed: July 7, 2021]

[7] Atzori L, Iera A, Morabito G. The Internet of Things: A survey. Computer Network. 2010;**54**:2787-2805

[8] Wang K, Zhao Y, Gangadhari RK, Li Z. Analyzing the adoption challenges of the internet of things (IoT) and artificial intelligence (AI) for Smart Cities in China. Sustainability. 2021;**13**:10983

[9] Huang Y-H, Chai C-W, Wu D-M, Chen L-G, Chao T-W, Chen Y-H. Research on strategy and technology of creating Smart City in Yilan with IoT Technology. In: 2021 International Conference on Electronic Communications, Internet of Things and Big Data (ICEIB). USA: IEEE Xplore. 2021. pp. 71-77. DOI: 10.1109/ICEIB53692.2021.9686373

[10] Chen P. The Construction of Indicators for Developing Yilan into a Smart Tourism Destination. Yilan, Taiwan: Fo Guang University; 2019

[11] Tsai C. Investigating the Creative City Development Strategy of YiLan from Intellectual Capital. New Taipei City, Taiwan: Fu Jen Catholic University; 2014

[12] Hsu N. The Development of Business Intelligence System for Regional Governance – A Case Study of Yi-Lan County. Taipei, Taiwan: National Chengchi University; 2016

[13] Lin J. Physical Planning of National Land Development for the Smart City - A Case Study in Yilan. Taipei, Taiwan: Ming Chuan University; 2019

[14] Yang W. Exploring the Development of Smart Cities through Public Private-People Partnership - Example of Taipei Smart City Project. Miaoli, Taiwan: National United University; 2021

[15] Lu Y. Research on 5G and IoT Application Development Strategies of Taiwanese Telecom Company for Smart City: A Case Study of F Company. Taipei, Taiwan: National Taiwan University of Science and Technology; 2021

[16] Xie Y. The Research on Key Factors of Innovation Governance of Smart Cities

in Various Countries. Changhua, Taiwan: National Changhua Normal University; 2020

[17] Planning Department, Yilan County, Taiwan. Smart Land Development Planning, 2022. Available from: https://planning.e-land.gov.tw/cp.aspx?n=707ED96D12B73148 [Accessed: June 21, 2022]

[18] Snyder H. Literature review as a research methodology: An overview and guidelines. Journal of Business Research. 2019;**104**:333-339

[19] Wohlin C, Höst M, Henningsson K. Empirical research methods in software engineering. In: Conradi R, Wang AI, editors. Empirical Methods and Studies in Software Engineering. Lecture Notes in Computer Science. Berlin, Heidelberg: Springer; 2003. p. 2765

[20] Chai C-W, Chao T-W, Huang Y-H. Research on the Application of Educational Robots for Distance Teaching in Rural Area for Post-COVID-19 - taking the Cultivation of Seed Talent Students of Indigenous Tribes as an Example, 2021 IEEE 4th International Conference on Knowledge Innovation and Invention (ICKII). 2021. pp. 223-229. DOI: 10.1109/ICKII51822.2021.9574777

Chapter 7

Communication Technologies and Their Contribution to Sustainable Smart Cities

Menachem Domb

Abstract

Sustainable smart cities (SSC) are becoming a reality as many develop their unique model of smart cities based on vast communication infrastructure. New technologies led to innovative ecosystems where transportation, logistics, maintenance, etc., are automated and accessed remotely. Information and communication coordinate their overall activities. Sensors embedded in these devices sense the environment to provide the required input. Together with artificial intelligence, machine learning, and deep learning, it enables them to facilitate effective decision-making. This chapter discusses the role of integrating technologies in smart cities, focusing on the information and communication aspects, challenges, limitations, and mitigation strategies related to the infrastructure, implementations, and best practices for attaining SSC. We propose a four-layered model covering the main aspects of incorporating communication technology within sustainable smart cities. It covers the basic physical level, providing guidelines for designing a smart city that supports the requirements of a proper communications infrastructure. The level above is the network level where we describe current communication networks and technologies. The rest two upper layers represent the software with integrated and embedded communication components. In summary, we conclude that communication technology is the key enabler of most of the activities performed in smart cities.

Keywords: smart city, sustainable, communications technologies, protocols, infrastructure, signal, wireless communications, urban area communication, signal attenuation, machine learning

1. Introduction

Typical SSCs are equipped with state-of-the-art technologies to support the new and modern lifestyle. Communication technologies are the primary enabler of most electronic interactions and associated operations. The European Parliament states, "A smart city seeks to address public issues via information and communication technology (ICT) solutions." The Japanese definition concentrates on energy, infrastructure, ICT, and lifestyle. Navigant Research [1] pointed out that investment in smart cities covers smart government, smart building, intelligent transport, innovative communications, and smart utilities. Wireless communication using electric signals is the core for accessibility and availability of communication everywhere within the city and at all times. The UN projected that 66% of the world's population will be urban by 2050, and cities consume most of the world's resources, such as 75% of the total energy. They will generate 80% of the greenhouse gases, causing adverse environmental effects. Smart cities with their inherent moderation and control of resource consumption are the ideal solution to address these challenges, population growth, deterioration of energy sources, environmental pollution, etc. The International Organization for Standardization (ISO) provides standards to assure a wide range of smart cities' quality, safety, and performance. Adherence to these standards benefits deploying, managing, and controlling smart cities. Implementing these standards requires embedding sensors within the involved devices and having these devices connected to a local network to establish inter-sensors communications using ICT. Nathali et al. [2] proposed a generic and universal bottom-up smart city architecture for real-world deployment. The architecture comprises four layers: sensing, transmission, data management, and application. Embedded communication means within each layer are critical and mandatory to ensure cooperation and synchronization among the various components of city sustainability.

ICT allows setting energy targets, observing, and enforcing them by deploying sensor networks covering primary energy consumption sources, such as municipal, industrial, hospitals, and citizens. A tool to identify optimal monitoring locations is available. In a case study, the ICT hotspots identified were heating systems, transport systems, and potential transformation of the buildings and roads enabled by ICT solutions. Studies show that a successful implementation requires the timing of ICT-related decisions in the planning process and the actor-networks needed to implement the ICT solutions and their management. The planning process has several decision points: the property owner, meta-network governance coordination, and traveler information systems. A flexible-work-hub solution case study revealed that mobility management systems encourage environment-friendly transport modes to reduce transport demand with minimum impact. All transportation means should be equipped with efficient navigation systems, and flexible work hubs should be located in local nodes closer to people's homes.

To provide a practical framework for this chapter, we propose a four-layer concept, an analogy to the OSI seven-layer for communications. **Figure 1** depicts the layers model, starting from the bottom with the physical city architecture layer, allowing electric and optical signals free of disturbances and delays. Then the network layer

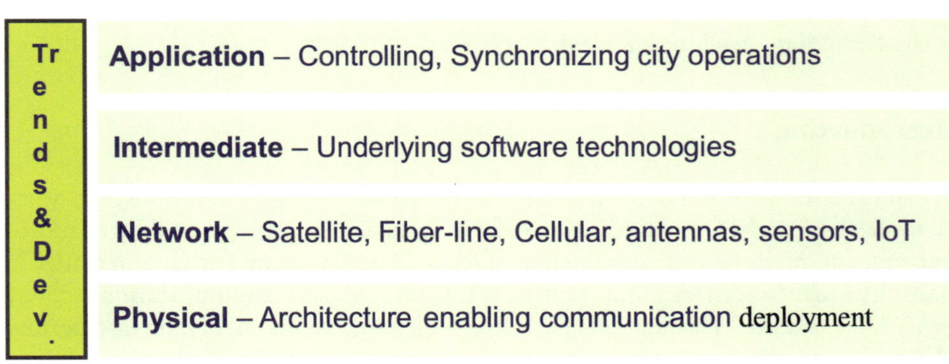

Figure 1.
The multi-layer framework for smart city communications.

where all the communication equipment, wireline, wireless, antennas, sensors, and IoT are optimally deployed. The intermediate layer contains the protocols and third-party software required to support the application layer and manage the lower-level devices. The upper layer has applications enabling the end-user to enjoy the benefits of a sustainable city, such as energy control and waste and pollution management. Each layer is improving over time; this is presented by the vertical column called trends and developments.

2. Trends and developments

The global smart cities market size is expected to expand from $1.226T in 2022 to $6.965T registering a CAGR of 24.2% by 2030. Navigant Research [1], based on 443 projects spanning 286 cities worldwide, will contribute nearly $1.7T to the global smart city technology market until 2030. Rapid urbanization led by government initiatives worldwide encourages sustainable and green technologies investment. Asia Pacific market seems to lead with a CAGR of 27.7%. Advanced cities use IoT to manage sustainable operations. For example, pollution, water, and healthcare. Endeavor Business Media announced the launch of smart-building technology embedding intelligence for new constructions and existing commercial buildings. This technological development reduces energy consumption. Major Asian mobile operators take many 5G deployments, and initiatives to resolve the problem of high bandwidth requirements are anticipated to drive the growth. The list of companies promoting smart cities shows that many leading communication vendors appear there, such as Cisco Systems, Inc., Ericsson, General Electric, Honeywell, IBM, Huawei Technologies, Siemens AG Telensa, Verizon, and Vodafone. The introduction of electric vehicles has been well accepted mainly due to their low pollution and modern look. However, it raised a new environmental issue of recharging stations and how to get rid of the big obsolete batteries.

To complement it, intelligent transportation systems (ITSs) [3] became decisive in minimizing congestion, pollution, and parking space. There is still a need for a closed monitoring system to prevent greenhouse gas emissions and promote efficient energy consumption, awareness, attraction, and broadcast decisions. Smart cities market report posted that the innovative utility section, the intelligent infrastructure, and the travel assistance segment are expected to grow at a CAGR of 22.9, 24.3, and 23.4%, respectively, over the forecast period. Endeavor Business Media, 06/21, announced the launch of intelligent construction technology combining smart communication components, reducing energy consumption. Waste management companies deploy sensor networks and data platforms to generate practical insights, route optimization, and analytics decisions. The growing adoption of new technologies in the smart ticketing market, RFID, QR code, BFSI, and healthcare offer smart solutions across sectors.

Businesses look for new ways to engage their customers, streamline operations, and generate revenue, and many are turning to wireless wide area network (WAN) technology. Wireless connectivity is now essential for enabling agile and secure connectivity of people, places, and things, beyond the reach and limitations of traditional wired network connections, managed wireless. The emergence of 5G, with its faster speed, lower latencies, and enhanced network capabilities, catalyzes wireless WAN adoption as businesses seek to make their WANs cellular simple and fiber-fast for true wireless flexibility. This solution provides businesses with the necessary secure and flexible wireless cellular connectivity to any number of fixed sites managed by network experts, helping organizations save time, money, and removing the burden of ongoing management or upfront infrastructure costs. Customers need an agile network that is quick to deploy, highly scalable, secure, and supports a broad WAN use case. They expect a plug-and-play, managed solution that enables simple and fast deployment of wireless connectivity when wired connections are unavailable, lack sufficient reliability, are too costly, only applicable to fixed locations, and require long lead times. Managed wireless WAN is designed to connect thousands of endpoints while providing end-users with fast and secure access to the cloud, datacenter applications, and the internet. It provides employees with safe and reliable access to be productive anywhere without relying on a network provider to deliver a circuit. 5G wireless edge devices offer connectivity, and plans for future additions to the service include support for in-vehicle and internet of things (IoT) use cases and the addition of enhanced routing and security features. Examples of particular use cases have a temporary connection at a branch site, pop-up store, or construction trailer, expanding to new locations, or using a permanent cellular connection as a failover or WAN link for an SD-WAN deployment. It extends the reach of the enterprise to remote areas https://www.computerweekly.com/news/252516487/European-employers-missing-the-opportunity-to-automate-processes-for-hybrid-work enabling innovative use cases.

3. Application layer: applications with embedded communication

Advanced communications enable the use of new services covering a variety of life indicators applications, such as shorter commute time, clean air, traffic control, street lighting, smart parking, gathering management, accelerated emergency response time, reduced healthcare costs, decreased water consumption, recycled waste, harmful emissions, sustainability, and other saving potential.

Figure 2 presents several key application types used in a typical SSC for managing, coordinating, synchronizing, and managing all city activities, such as advanced metering of water, electricity, and gas consumption control. Real-time metering of measurable elements, anomaly detection, alert systems, sensor-data collection, machine learning, deep learning methods, and big data analysis. The expected impact is efficient, balanced, cost-efficient, reliable, secured, improved power consumption, low air pollution, and tight coordination among city sectors, such as energy, transportation, water supply, healthcare, education, and culture. In parallel, privacy and security issues are handled, and centralized IoT applications for cost reduction and energy saving of LED lighting controls. Applications for managing surveillance cameras, environmental sensors, electronic billboards, charging stations, WiFi coverage, and smart transit systems reduce cost, improve safety, and routing management improve user experience,

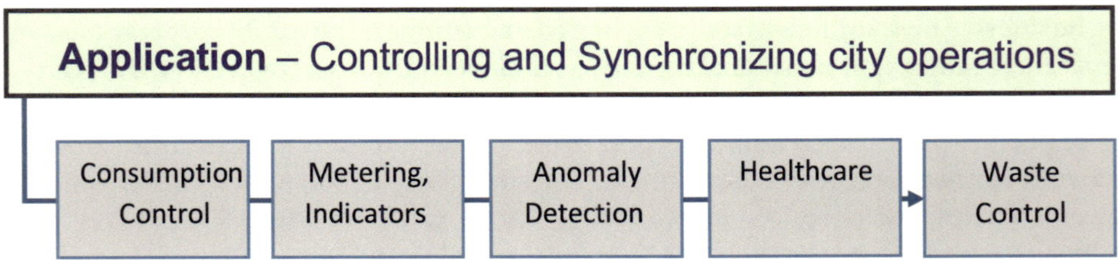

Figure 2.
The application layer detailed examples.

onboard WiFi reduce congestion, provide clean air quality, and priority access management. Other applications, sensing flow rates, tank pressure, water levels, remote management solution, monitoring or the components of the IoT, including a range of radio devices, system on modules (SOMs), sensors, water management applications, gateway to supply the connectivity for a range of application needs, water treatment solutions, evaluate critical environmental data, such as groundwater analytics providing recommendations to customers. Remote monitoring and management solution offer hidden visibility of equipment and customers usage of chemicals.

A collection of vast applications arsenal is the core enabler of SSC. Following are estimated global market values per application type: [4]. Smart metering for the electricity, gas, and water, market is estimated to reach $39B by 2027. The smart lighting market will grow at a CAGR of 18% to $31B by 2025. Intelligent electric vehicle (EV) charging market is expected to reach $70B by 2026. The solar photovoltaic (PV) market is estimated to grow by 5% annually, reaching $185B.

4. Intermediate layer: communication management

This layer provides the underlying generic technologies required by the application layer to operate, giving new ideas and capabilities, and empowering the software intelligence to a leading position in the software domain. **Figure 3** depicts state-of-the-art technologies enabling AI and other libraries to enrich the applications in the first-layer. The API library contains various generic software components the application layer uses. Big data is another component having a warehouse of data collected over a long period enriched with related market data. Data mining, AI, and BI use this data to identify data patterns, rules, and exciting insights. Machine learning (ML) and deep learning (DL) are two modern tools that are able to learn some insights by processing a given training data. These insights are then used to extrapolate and predict the behavior of the system results. Cloud computing transforms computer-owned usage into services without owning the computer environment. It is disconnecting computer services from the organization's site. Consequently, the software can be accessed anywhere and anytime free of maintenance, which is an excellent advantage for a smart city. Cyber security is a comprehensive solution to secure the entire system from cyber-attacks.

The following are typical qualifications representing SSCs [4], as follows: Technological provision, environmental, social, economic sustainability, economic and social development, air quality, energy transition toward renewables, quality of living, waste per population, water sustainability, human infrastructure & networked markets, ESG performance, and smart city ecosystem. Some of the cities provided data regarding their status. One city deployed over 20,000 sensors for capturing

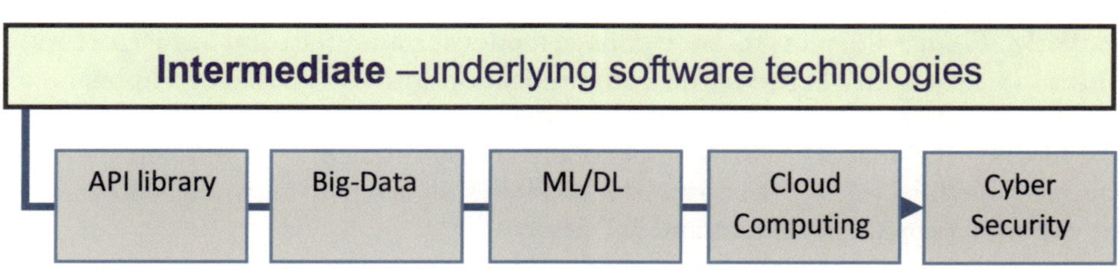

Figure 3.
The intermediate layer detailed examples.

temperature, air quality, mobility data, lighting, noise, and climate. Another city implemented pollution-monitoring sensors and educational campaigns. Some cities stated that all new buildings are built with intelligent controls, low-energy heating, and digitized mobility using accessible WiFi in 755 public spaces. More options are wired bike-sharing, electrical vehicle plug-in spots, activated video feeds in busy intersections to smooth traffic, renewable energy, sustainable mass transit, $70B in total startup valuations, 100 accelerators, incubators, and co-working spaces, using 100% renewable power, implementing real-time meter sensors, reducing emissions from daily commuting by sharing, and deploying sensors for heating, cooling, and lighting based on occupancy. Distribute to the public smart-mobile applications, measure and optimize biogas, energy efficiency, heating and cooling, smart grids, and consider electric buses and green energy systems.

Following are typical declarations of existing smart cities. The city goals are clean air, biodiversity, low carbon, green transportation, waste reduction, artificial intelligence (AI), blockchain, internet of things (IoT), quantum computing helping in their intelligence journey, and cutting 40% of CO_2 emissions by 2025. Becoming the leader in smart and sustainable building solutions. Through the $37N green building masterplan, make 80% of the city buildings eco-friendly by 2030, earn 80% of new buildings super low energy (SLE), and achieve an 80% improvement in energy efficiency for green buildings. Becoming a climate-friendly by 2040. Any services that can be digitized will be digitized. Become the world's first carbon-neutral city by 2025, becoming fossil-fuel-free by 2050.

Some of the recorded achievements are: reducing carbon emissions by 25,000 tons, saving $9.5M, decreasing the electricity consumption of public buildings by 7.8%, reducing overall carbon footprint by 35%, recover 1.64 million tons of municipal solid waste, reducing emissions by about 18,000 T/year, comparable to the electricity use of 4000 residents, Over a third of all transportation, fossil-fuel consumption has been removed through sustainable transport alone, a reduction of 90,000 tons of greenhouse gas emissions each year.

5. Network layer: city wireline, wireless 5G, WiFi7, antennas, sensors, and IoT

Several publications define the requirements for qualifying a city as an SSC. In all of these publications, we realize that communication and sensors are the key enablers of smart cities. Internet connectivity is crucial for smart cities as almost all activities are via messaging. High capacity, high-speed, efficient, and effective internet connection is a key to achieving the smart cities vision. It complies with the forecast that by 2024, more than 23B devices will be connected to cellular networks. It is possible with high-speed internet connectivity associated with local communication networks. **Figure 4** depicts recent trends in modern communication infrastructure, which can cope with a high-volume communication activity. The first component is satellite communications, which is undergoing significant development by SpaceX. The second is cellular 5G, which supports a new magnitude of transmission speed and volume. The third refers to a substantial WiFi version, a newly expanded gateway, and the exploding spread of sensors and IoT devices.

During the past few years, we are evident the intensive launch of more than 2000 small satellites to the LEO by SpaceX, creating a network of satellites communicating with each other via laser beams. The communication with the earth is by electronic

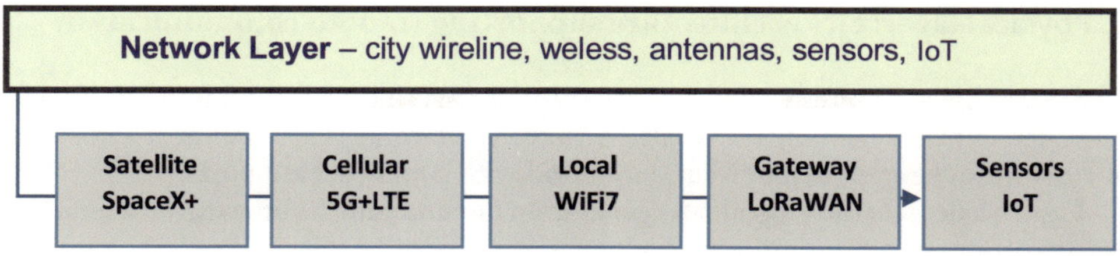

Figure 4.
The network layer details.

signals transmitted toward ground stations for further distribution via wireline and cellular networks. It is the ultimate achievement of satellite communications.

The evolving spread of 5G provides ultra-fast internet, low latency, and improved reliability. It is the ultimate solution that copes with the expected wireless traffic. 5G network's speed is 10–30 Gbps, which is 100 times faster than 4G; the capacity is 1000 Gbps/km^2 area spectral in dense urban environments, 1000 times more than 4G. It decreases energy consumption by 10% times the higher battery life of associated devices and five times lessened end-to-end delay. 5G integrates with long term evolution (LTE) and WiFi to give all-inclusive high-rate coverage and a seamless user experience. 5G networks have a latency rate of 1 ms vs. 40–50 milliseconds in 4G. 5G networks allow smoother handling of spikes and better network traffic optimization than 4G. Lower power consumption and enhanced capacity and speed are part of sustainability.

WiFi wireless communications transport most wireless traffic in enterprises, public and residential environments cost-effectively and continue improving the efficiency in using precious spectrum resources. The new version 7 is to be released in 2024. It is a significant enhancement of WiFi 6. It is more flexible and efficient, supports 16 streams, has a channel size of 640 MHz, has a data rate of 46 Gbps, has lower latency, and uses network and spectrum resources. WiFi 7 integrates well with 5G and 3GPP-based 5G and other standard communication devices and protocols. It supports distributed and cloud architectures, virtualization, and digitalization in the emerging private wireless networks (PWN). Wi-Fi-7 supports applications that require deterministic latency, high reliability, quality of service (QoS), IoT, IIoT, and video-based applications, such as surveillance, remote control, gaming, AV/VR, smart-home services, and more. WiFi deployment provides communication services that save unnecessary wiring, energy, transportation, and contribute to sustainability.

The evolving new services generated for smart cities require numerous sensors connected to new types of wireless communication networks that meet the specific requirements of smart city needs. Sensors [5] interactions require the transfer of small data packages, energy efficiency, the ability to connect devices in remote areas, a high degree of data protection, and interoperability. Connected end devices must operate for a long time, powered by an embedded battery with no connection to the grid. Terleev et al. [6] recommend LoRaWAN as the best gateway for machine-to-machine communication technology. According to experiments, the coverage area of the LoRaWAN gateway is 1500 m, which is fine for a smart city.

The last component required to complement the network layer is IoT, the internat of things, enabling the data collection from the system endpoints, the sensors, and vice versa, transmitting messages from the system toward an IoT device and among IoT devices.

6. Physical layer: city architecture supporting smooth communication

Smart city communication infrastructure supports intra-city and internet interactions. It comprises wireline and wireless mixture networks. The wireline is a network of fiberoptic channels deployed underground with connected antennas.

Figure 5 depicts the typical new generation of communication hardware required to support modern communication services. It includes satellite and cellular antennas, underground fiberoptics wiring, and the construction materials impacting the electric signals. The number of antennas, location, signal strength, and height depend on the city's population density. The wireless portion comprises signals from satellites intercepted by the corresponding antennas and signals broadcasted by the cellular antennas and captured by the mobile phones located within the antenna's spectrum. Wireless signals are disturbed by physical obstacles, such as buildings and other constructions. Therefore, city streets, buildings architecture plans, and used materials should consider optimal deployment of the wireline fiberoptics and the corresponding antenna locations to ensure smooth communication at minimum interference. For example, the building material and the estimated data transmission load apply the suitable communications infrastructure or determine the building's fabric. We provide the knowledge and guidelines for selecting the appropriate communications technologies fitting the specific SSC's attributes and vice versa.

Electric waves are the core of wireless digital communication at free space and ground contacts. However, the transmitted waves are exposed to obstacles, such as rain, dust, topography, urban surface, and magnetic forces, causing signal attenuation and degrading the transferred signal quality up to data loss. To overcome it, we may request the transmission of stronger signals, which increases the power consumption and shortens the transmitting satellite's lifespan. Hence, we propose a machine-learning based model, which predicts the proper signal strength and the correct transmission time, having a high probability of reaching the intercepting antenna on the ground. The model analyses the two path sections, from the satellite to free space close to the ground and then to the ground station. We trained our ML system using training data from the genesis satellite. Experiment results show our system's high accuracy level for frequencies ranging from 2 to 72 GHz.

Several papers cope with the same problem. Some proposed solutions are limited to a geographic region where minimal rain and dust, while others are limited to low frequencies [7]. Analyzed satellite data to discover the elements causing a signal loss in urban environments [8, 9]. Correlate signal loss and construction material [10, 11]. Present materials measurements of low frequencies [12, 13]. Focus on the receiver's position and height disruption inside a building. Entry loss for 2 GHz is reported in [14–18]. Discuss signal spread within facilities and [19] calculate the spread delay as a function of elevation and angle. In [20], a new path loss model and [21] present attenuation differences

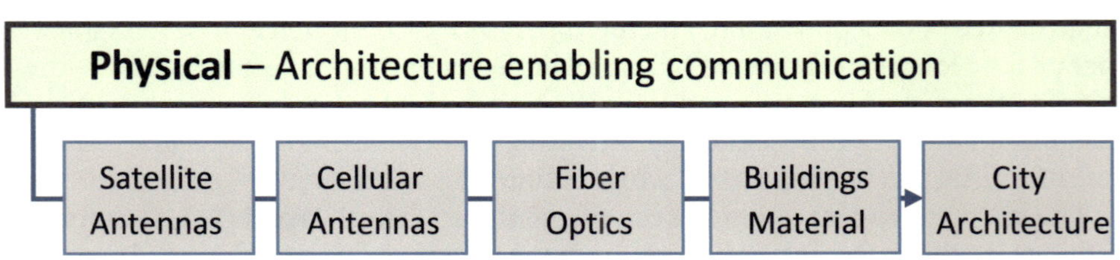

Figure 5.
The physical layer details.

for indoor textures and materials. We selected a concrete building with a horn antenna with vertical polarization. The signal generator, the transmitter, and the receiver are agilent. We use recurrent neural networks (RNN), and MLNNs have a temporal dimension. The MLNN system builds, loads the neural network, trains, and tests the input data. Then it analyzes the data, starting with the input samples reduced in two stages to one instance, identifying the optimal converged parameters. The training results resemble the accurate results. In summary, we provide SSC designers with a tool to determine the optimized materials and positioning of communication equipment so that signal strength will remain effective until it reaches the targeted antennas and mobile devices.

7. Summary and conclusions

This chapter discusses the role and contribution of communication technologies to sustainable smart cities. For a detailed analysis, we divided the subject into four cumulative aspects piled into four layers, starting with the bottom physical layer up to the applications developed to manage, coordinate, and synchronize the activities required to maintain sustainability in smart cities. This study shows the profound necessity of incorporating communication technology into the management and control mechanisms used to assert sustainable smart cities. Based on the detailed content of the chapter sections, it is clear that everything related to management, control, and interaction

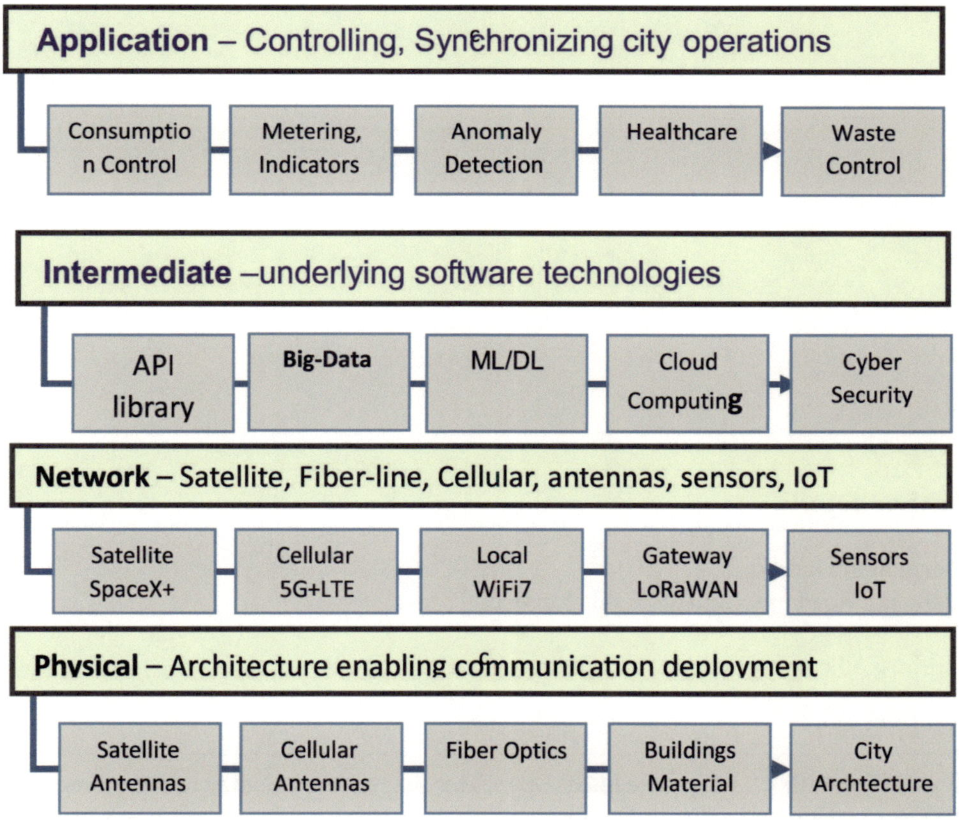

Figure 6.
The multi-layer detailed framework for smart city communications.

among key players of smart cities require communications infrastructure to enable its operations. **Figure 6** encapsulates the complete view of the chapter content and its details. Since SSC is still in its beginning stage and still evolving, this chapter may be updated soon to capture the near future developments in this advanced domain.

Conflict of interest

The authors declare no conflict of interest.

Author details

Menachem Domb
Ashkelon Academy College, Ashkelon, Israel

*Address all correspondence to: dombmnc@edu.aac.ac.il

References

[1] Navigant Research's Smart City Tracker 2Q19 Highlights. June 20 2019

[2] Nathali B, Bhagya S, Silva N, Khan M, Jung C, Jung C. Urban planning and smart city decision management empowered by real-time data processing using big data analytics. Sensors. 2018;**18**(9):2994. DOI: 10.3390/s18092994

[3] Haferkamp M, et al. Radio-based traffic flow detection and vehicle classification for future smart cities. In: 2017 IEEE 85th Vehicular Technology Conference (VTC Spring). 2017. pp. 1-5. DOI: 10.1109/VTCSpring.2017.8108633

[4] https://www.silabs.com/applications/industrial-and-commercial/smart-cities?source=Media&detail=Google_Ads&cid=pad-gos-all-111921&gclid=Cj0KCQjwwJuVBhCAARIsAOPwGAQHA29BqXdFGWJiNB5ZzbDiOU9oAkuP9z_AR7bu-Q1p88uws-Mi9LEaAioTEALw_wcB

[5] Stone W. Electromagnetic Signal Attenuation in Construction Materials. Gaithersburg, MD, USA: National Institute of Standards and Technology; 1997

[6] Terleev AV, Khalturin AA, Shpenst VA. LoRaWAN gateway coverage evaluation for smart city applications. In: 2021 3rd International Youth Conference on Radio Electronics, Electrical and Power Engineering (REEPE). 2021. pp. 1-4. DOI: 10.1109/REEPE51337.2021.9388004

[7] Al-Hourani A, Guvenc I. On modeling satellite-to-ground path-loss in Urban environments. IEEE Communications Letters. 2021;**25**(3):696-700. DOI: 10.1109/LCOMM.2020.3037351

[8] ITU-R Rec. P.1411-3: Propagation data and prediction methods for planning short-range outdoor radio communication systems and radio local area networks in the frequency range 300 MHz to 100 GHz. 2005

[9] Perez-Fontan F et al. Building entry loss and delay spread measurements on a simulated HAP-to-indoor link at S-band. EURASIP Journal on Wireless Communications and Networking. 2008;**2008**:1-6

[10] Baker-Jarvis J, Janezic MD, Riddle BF, et al. Measuring the Permittivity and Permeability of Lossy Materials: Solids Liquids, Building Material, and Negative Index Materials. Gaithersburg, MD, USA: National Institute of Standards and Technology; 2005

[11] Stavrou S, Saunders SR. Review of constitutive parameters of building materials. In: Proceedings of the 12th International Conference on Antennas and Propagation (ICAP '03); March–April 2003; Exeter, UK. pp. 211-215

[12] De Toledo AF, Turkmani AMD, Parsons JD. Estimating coverage of radio transmission into and within buildings at 900, 1800, and 2300 MHz. IEEE Personal Communications. 1998;**5**(2):40-47. DOI: 10.1109/98.667944

[13] Martijn EFT, Herben MHAJ. Characterization of radio wave propagation into buildings at 1800 MHz. IEEE Antennas and Wireless Propagation Letters. 2003;**2**(1):122-125

[14] Vogel WJ, Torrence GW. Propagation measurements for satellite radio reception inside buildings. IEEE Transactions on Antennas and Propagation. 1993;**41**(7):954-961. DOI: 10.1109/8.237628

[15] Oestges C, Paulraj AJ. Propagation into buildings for broadband wireless access. IEEE Transactions on Vehicular Technology. 2004;**53**(2):521-526. DOI: 10.1109/TVT.2004.823546

[16] Devasirvatham DMJ, Krain MJ, Rappaport DA, Banerjee C. Radio propagation measurements at 850 MHz, 1.7 GHz, and 4 GHz inside two different office buildings. Electronics Letters. 1990;**26**(7):445-447. DOI: 10.1049/el:19900289

[17] Nobles P, Halsall F. Delay spread measurements within a building at 2 GHz, 5 GHz, and 17 GHz. In: Proceedings of the IEE Colloquium on Propagation Aspects of Future Mobile Systems; October 1996; London, UK. pp. 8/1-8/6

[18] Report ITU-R. Compilation of measurement data relating to building entry loss. May 2015

[19] https://www.disruptive-technologies.com/blog/the-top-20-sustainable-smart-cities-in-the-world

[20] Plets D et al. Simple indoor path loss prediction algorithm and validation in a living lab setting. Wireless Personal Communications. 2013;**68**(3):535-552

[21] Micheli D, Santoni F, Delfini A. Measurement of electromagnetic field attenuation by building walls in the mobile phone and satellite navigation frequency bands. IEEE Antennas and Wireless Propagation Letters. 2014;**14**:1. DOI: 10.1109/LAWP.2014.2376811

Chapter 8

Smart City Serious Game Based on Features Selection

Fachrul Kurniawan, Supeno Mardi Susiki Nugroho, Mochamad Hariadi, Isdaryanto Iskandar and Prita Dewi Basoeki

Abstract

In general, the smart city concept integrates ICT devices to collect various types of data related to urban life. The data will be used to manage and improve assets and operations in the city. In determining a smart city, there are eight key aspects (features) as a consideration. However, implementing eight smart city features will require many resources, including time, cost, and human resources. Therefore, this paper aims to develop a scenario and method of features and smart city parameters establishment in a serious game. This game consists of visualization, storyline, and gameplay as a learning platform to increase players' understanding of the smart city concepts. The development was started by collecting feature data of each city. The collected data will be discovered for the relationship between its features using differential equations. Then, data processing results will be classified and used as a basis for serious games. This study's product is an easy-to-use game that can simulate planning and constructing a smart city. This game is intended for the city government or the mayor as a critical role in city development. It will ease the user to understand the concept of the smart city based on the city characteristics. Furthermore, this serious game will provide feedback and recommendations on aspects that need to be improved in the city. It can be basic knowledge to make an actual decision and policies in smart city development.

Keywords: city development, serious games, smart city

1. Introduction

The challenges to implementing a smart city concept in developing countries are technology adaptation, government operational management, and lack of clear strategy [1]. As one of the developing countries in South East Asia, Indonesia has potential for smart city development [2, 3]. However, this archipelago state implements smart city is difficult. Each city has its characteristics, so that implementation cannot be equated. In general, these cities are classified based on the number of population, which are small city (<100,000 population), medium city (100,000–500,000 population), big city (500,000–1,000,000 population), and metropolis (>1,000,000 population).

There are several considerations to determine a smart city, such as a feature of the smart city. In its application, there are eight main features that each feature has several parameters. The factor will decide the level of a city, whether it is categorized as a smart city or not [4, 5]. It is tough to implement all the eight smart city features because it requires a lot of resources. Therefore, it needs a platform that can simulate planning and constructing an ideal smart city. One simulation model is in the form of a serious game [6, 7].

In this case, the use of serious games has several advantages. This kind of game emphasized a specific purpose such as education, scientific exploration, city management, etc. [8–10]. The smart city serious game aims to simulate the implementation of the smart city based on features. Based on the simulation through the game, it can be a consideration as a problem solving related to the smart city issues. Furthermore, using these serious games helps to save costs, human resources, and time.

According to the researcher's project, a smart city-based development project provides many data from the smart operation room. It records issues around the city, such as delay in the development projects and the determination of bandwidth network technology needed to integrate CCTV cameras and operation rooms. This phenomenon encourages researchers to use a game-based method as part of problem-solving to find the most effective development system [11, 12]. One of the expected outputs from the serious game is the clarification optimization process that can be implemented in city development [13].

The game will provide feedback, knowledge, and specific result related to the city's problem. It will make a player focus on the issues and create a solution in starting the development using the smart city system [6, 14]. The scenario of these games is adjusted to the city's problems covered in eight features smart city [4, 15]. In the next section, it will describe the formulation of smart city features and parameters.

2. Smart city features and parameters

The formulation of smart city features is based on current development in a developed country [16–18]. Those features are smart health, smart education, smart mobility, smart energy, smart government, and smart technology. However, there are

Smart city features	Parameters
Smart health (HE)	Medical team, disease sufferer, blood donors
Smart education (ED)	School, university, college graduate
Smart mobility (MO)	Vehicle, mobility, traffic accident
Smart energy (EN)	Electricity, water
Smart government (GO)	Demography, state civil apparatus, regional income
Smart technology (TE)	Social media, internet, smartphone user
Smart infrastructure (IN)	Industry, agriculture, farm, fishery, school, hotel, hospital, worship place, government offices, fiber optic
Smart people (PE)	Population density, urbanization, birth rate, death rate

Table 1.
Smart city features and each parameter.

two additional features based on characteristics and differences between developed and developing countries: smart infrastructure and smart people. Furthermore, there are some parameters to measure each feature value. **Table 1** presents the smart city features and each parameter. Based on **Table 1**, those parameters are collected from the Indonesian central bureau of statistics. The data of parameters will be processed and resulting in a main value of features. In general, the feature data will be compared end evaluated with the standard of a smart city. The information of current city condition becomes a fundamental consideration to improve the low sector.

This feature formulation is essential as preparation of scenario development in the smart city serious game. The formulation process of feature value is based on differential equations. The final score for each feature is 0.125, so that the overall value is 1 or 100%. The results will be used to give weight to the serious game scenario, and then it will be classified into three categories (not ready, standard, smart city). This classification will determine the game's feedback on a smart city serious game scenario.

3. Proposed method of the smart city serious game

The research process begins with identifying data based on several previous studies related to the smart city. This identification produced the required features and parameters that are used for smart city determination. In the previous section, it is known that eight features were identified in this study. After that, the collected data will be processed using a differential equation and classification algorithm to develop a whole smart city serious game (visualization, storyline, and gameplay). **Figure 1** presents the research flow of smart city serious game development.

Data resources become an essential thing in this study. It was collected from the Indonesian central bureau of statistics within the last 3 years. It is crucial to use credible data as a basis for future experiments. For example, **Table 2** presents the energy feature data that consists of two parameters: electricity and water. Researchers are also participating in smart city planning projects to enrich data identification and explore related problems.

After the data identification and collection, it continues with the data modeling using a differential equation. This step aims to find out the difference of smart city feature data and determine each feature's urgency level. The use of differential equations also aims to determine the weights in the game process. Also, the slope one algorithm is used to discover data comparisons in each city. It can provide a recommended formula for smart city development. Next, the data classification is calculated by the learning algorithm. This algorithm will classify the data into three categories (not smart, standard, and smart). This classification process is important because it is the feedback that the serious game gives to the players.

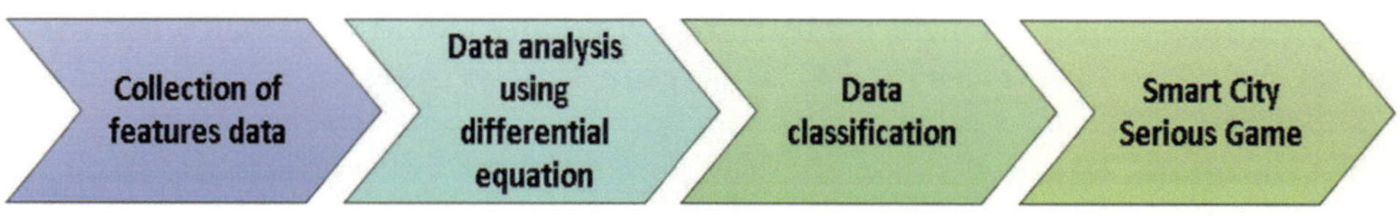

Figure 1.
Research flow.

City	Parameter of energy features						
	Electricity			Water			Smart energy
	Customer	Population	ax	Customer	Population	ax	
Jakarta	4,819,168	10,075,310	0.478	813,356	10,075,310	0.081	0.559
Cirebon	132,686	305,899	0.433	51,516	305,899	0.168	0.602
Pasuruan	76,169	193,329	0.393	224,210	193,329	1.159	1.554
Surabaya	1,043,501	2,765,487	0.377	526,688	2,765,487	0.191	0.568
Kediri	78,321	293,282	0.267	13,900	293,282	0.047	0.314

Table 2.
Data of energy features and its parameter.

4. Result and discussion

4.1 Scenario design of smart city serious game

The purpose of scenario design development is to identify a required process. In this study, the fidelity aspect means that simulation cases are based on the actual data. Furthermore, this game trains the process of identifying cities and their problems, especially for the local city government. **Figure 2** presents the scenario of a serious game. **Figure 1** started with the type of city based on the population number (small city, medium city, big city, and metropolis). From the city classification, players are directed to understand the differences from each city in Indonesia.

After understanding the type of a city, players are required to know about the concept of a smart city through its features and parameters. Players will be invited to study the smart city components, which consist of features and parameters data. By introducing the smart city components, players can imagine the concept of the game. Further understanding of the smart city concept can be reached after players completed the parameters of each feature. After that, it will be processed, and the results will be given in the form of a graph. It describes and concludes the gameplay, also provides a detailed description of each feature visualized in the graphic.

4.2 Main principles of smart city serious game

There are three essentials principles in developing smart city serious games: learn, rule, and play. It aims to make the players reach as much benefit as possible, and they would completely understand the concept of smart city development.

The learning principle is related to the type of city. By learning the city type, the player will understand the characteristics and issues of each city. Then, it would give feedback on development projects that were supposed to be prioritized [14, 19]. With an entertaining concept, a different learning experience would be gained when the players understand the issues [12, 14, 20]. **Figure 3** shows the learning process of city data identification.

Furthermore, **Figure 4** shows that the population of a city affects the city's type and the problems faced by the city. Thus, the score and development treatment would be different, depending on the type and level of the city's problems. This learning experience through the game will affect knowledge improvement. **Figure 5** shows that the level of understanding was started with level 1 with

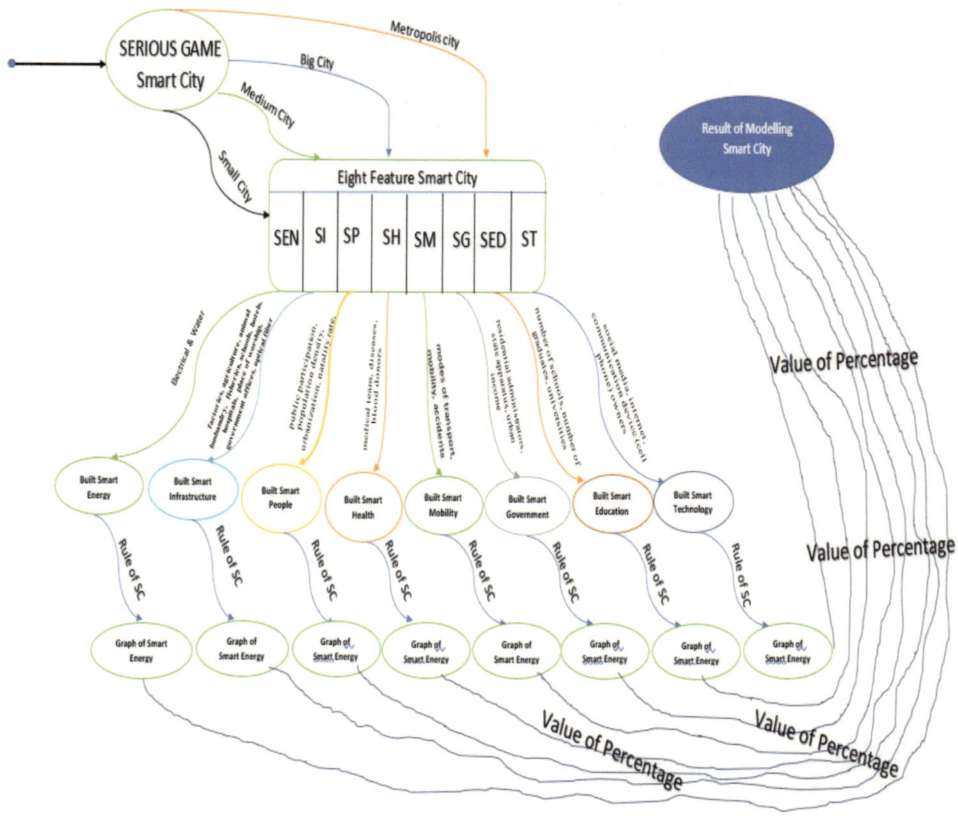

Figure 2.
The scenario of smart city serious game.

limited data, 10%. As the level increases to level 2 or 3, the cities were getting bigger and more complex problems. It allowed the players to get more experience and improve their understanding.

In a serious game, rules are necessary as a basic guideline related to the actual case. The serious game rule must be in accordance with the real facts, and this interaction would be relevant to the real case. Each feature has been given a weight based on its priority level that becomes a rule for the smart city serious game. **Figure 6** shows the rule function control in serious game development.

According to **Figure 6**, the rule function that formed a set of data relevant to the standard and actual conditions. Then, it was used to set training data to provide scores for the smart city system's eight features. Each rule had an objective in training: to get the general description of the training to collect relevant conditions and result in actions that followed the training guidelines. This rule was highly important and affected the users' knowledge of the cities' problems, which were seen from the scores of the parameters.

The last principle of smart city serious game is play. Raw data collected from the survey were input to be processed using a clustering system and classified before the training process. The game did not use levels to show acceleration. Instead, it required the players to prove input based on the types of cities, and thus an optimal score would be gained at the end of the game. The game would provide recommendations to players on required actions to solve a similar problem.

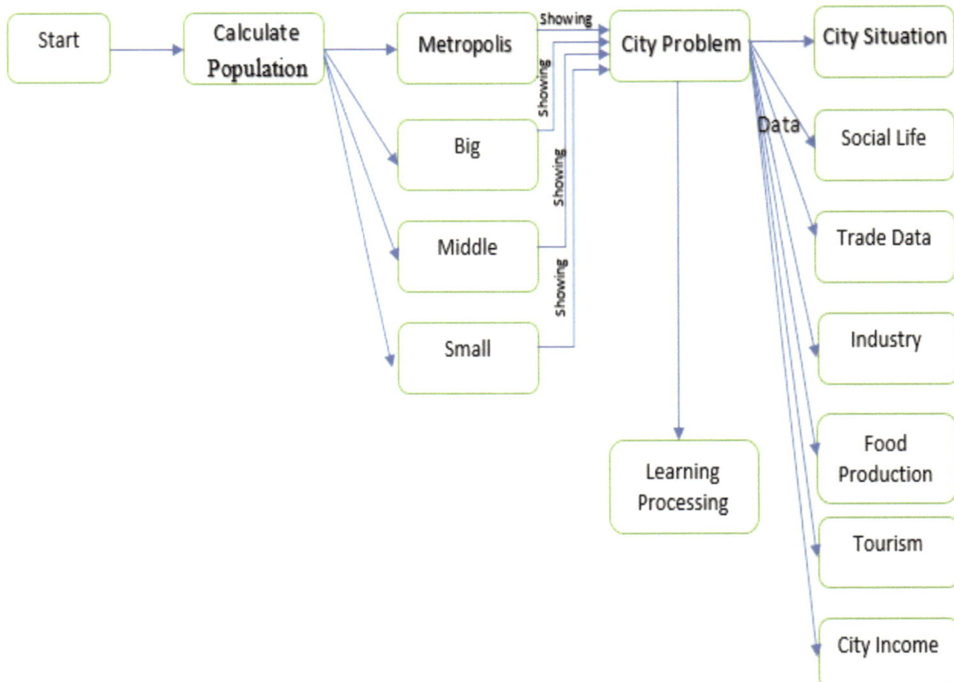

Figure 3.
The learning process for city identification.

Figure 7 shows the indicator input in accordance with the game level. It started with the small city category and increasing along with the higher level. The level of the game is increasing while the player completes and understands the current level. After that, the player can play at a higher level with the higher type of city and face more complex problems.

4.3 Implementation smart city features in the serious game

As we mentioned before, the game begins by selecting the city's type based on the population number. It is important to understand the players who act as city mayors to consider the characteristics and limitations. **Figure 8** previews the city selection page in the game.

After the city type selection, the game will preview the features and parameters of the city. The player has to input the number according to the actual data. The input of real data is vital so that the system will provide a calculation and feedback as accurately as possible. It can be basic knowledge to develop the current city condition. For example, **Figure 9** previews the input page of the smart energy feature in the small city. Based on the preview, it can be seen that the smart energy feature consists of two parameters (electrical and water).

The category of the city could be seen from the number of populations shown and from the eight features of the smart city as well as their parameters. **Figure 10** presents another feature (smart infrastructure) with its parameters.

After the player completes all the features and parameters, the data will be processed using the impact factor score determined in the system. The game's result

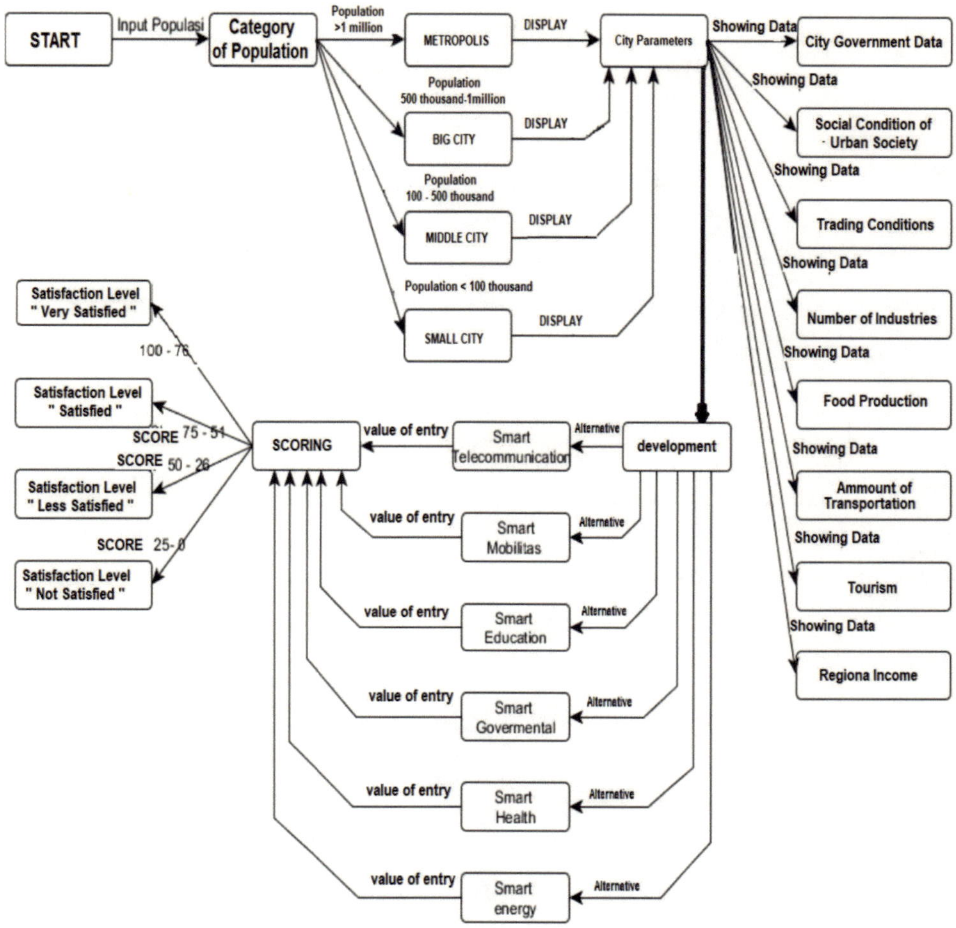

Figure 4.
Diagram of learning data.

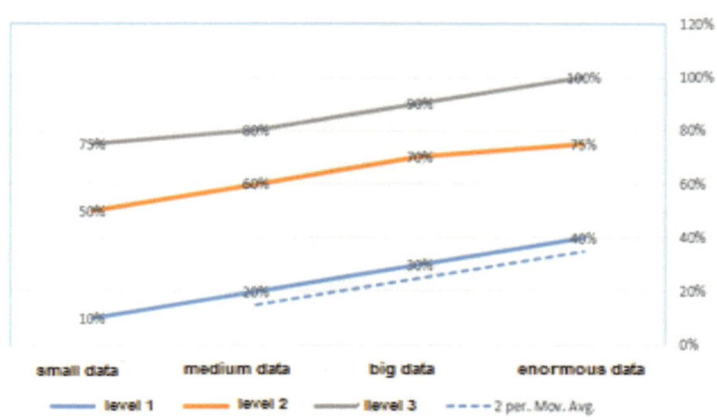

Figure 5.
Level of game understanding.

concludes that the mayor's policies are around 34% of development. It means that the development of the smart city is relatively low. **Figure 11** presents the result and recommendation page of the game.

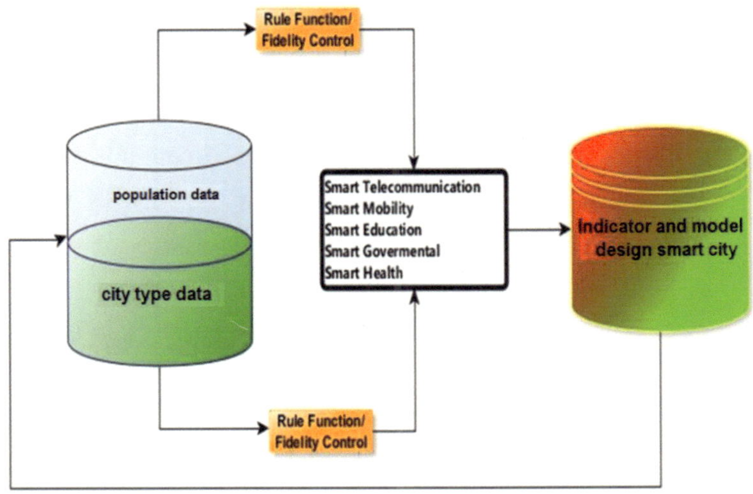

Figure 6.
Rule function (fidelity) in smart city serious game.

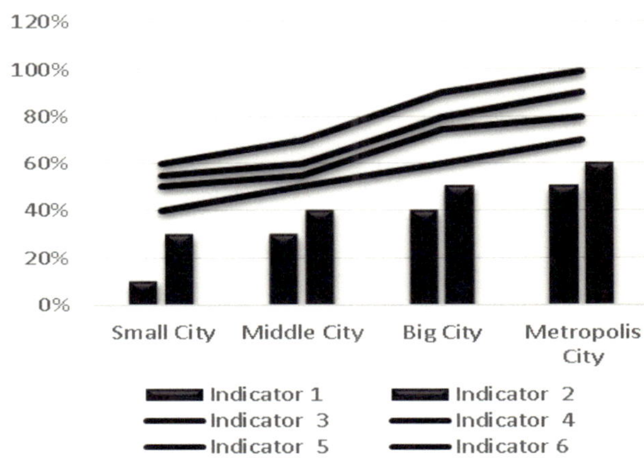

Figure 7.
The indicator of data input based on the game level.

Furthermore, detailed info will be provided in the graph that shows the city conditions. It shows the value of each smart city feature. It is used to evaluate policies and optimize values lacking in smart city features to make improvements in the next game. This feedback and recommendations make a player more focus on areas that need improvement to meet the smart city's standard.

The score of the game estimates the mayor's level of understanding of the city's problems. It could help to improve the level of understanding of smart city-based development programs. The smart city game could be useful if used continuously until it achieved a standardized score as intended by the player (city mayor). It could eventually improve the smart city system's knowledge and provide many considerations in determining developing cities' policies.

Moreover, the use of smart city serious games has many advantages [21, 22]. It helps the player to simulate that appropriate with the actual city cases. It can save time, costs, and resources before implementing the real improvement. It is also used

Figure 8.
Selection of the type of the city.

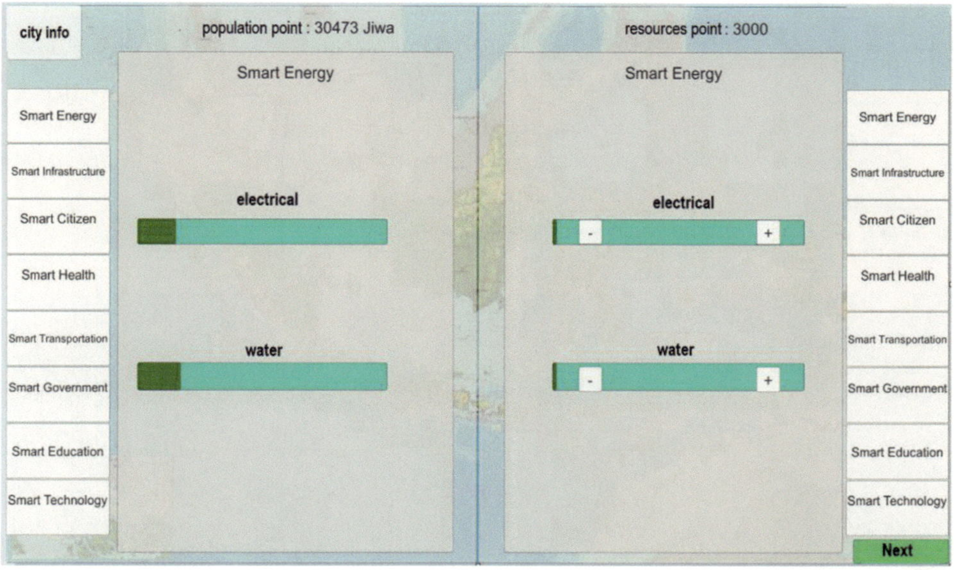

Figure 9.
Input page of smart energy feature in the small city.

to minimize a potential failure due to the lack of understanding by the city mayor [23]. Furthermore, the smart city serious game is expected to help an equitable city development, especially in developing countries.

5. Conclusion

The differences between developing and the developed country become a factor in the development of the city. Infrastructure and human resources are the two essential differences of it. Therefore, more considerations are used so the development can be implemented properly. The serious game could be considered a training and learning platform to decide and policies in actual city development. In smart cities, the game could help minimize failure for the city mayors in implementing development programs. Also, it would save the required time, costs, and resources

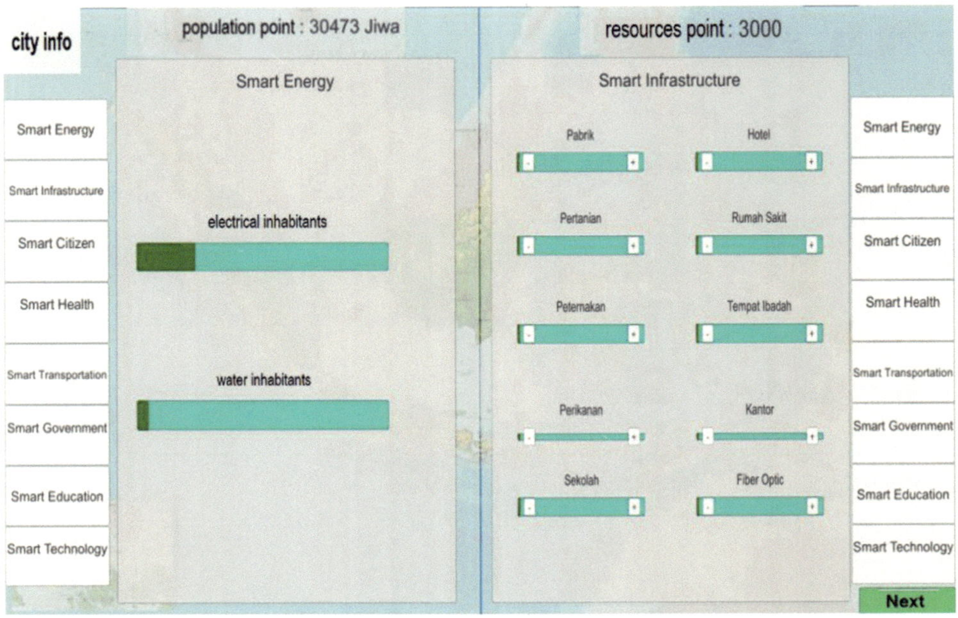

Figure 10.
The input of smart energy and smart infrastructure parameters.

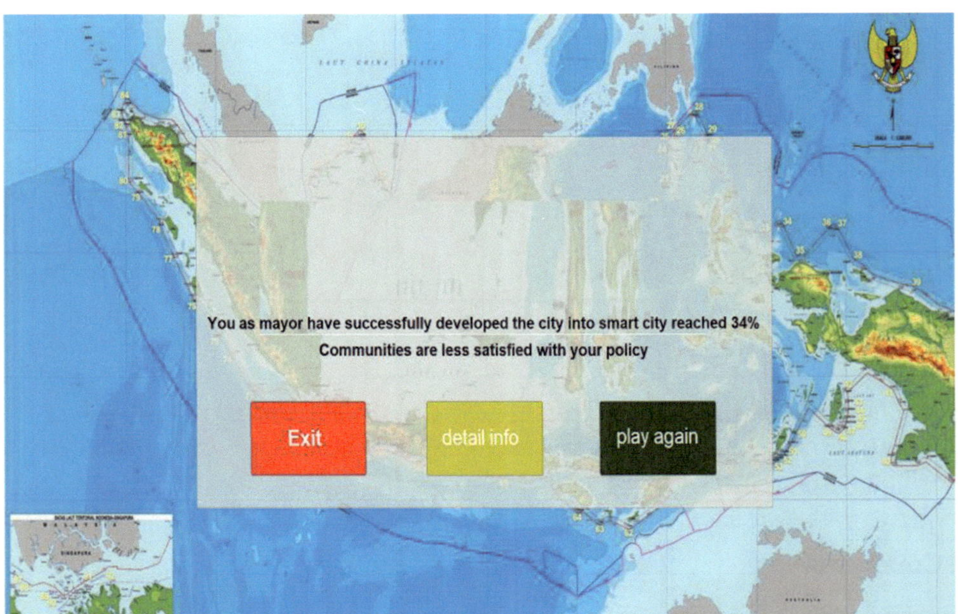

Figure 11.
The result and recommendation page.

before implementing the city development. Features and parameters of the city are an indicator to determine the smart city level. Thus, it must be customized with the characteristics of the country or the city. The formulation of features and parameters is important to help each city to reach the potential improvement. The three essentials principles: learning, rule, and play, allow the player to understand the characteristics and complexity problem in the city. It makes a player, especially a city mayor, increases their understanding of the smart city concept.

A serious game scenario design methodology is still highly challenging for further study since it is often not relevant to the actual conditions. It is needed and optimization to produce an adequate configuration. The development related to big data and business intelligence is also considered in the next study. It will help the government to make a policy and solution based on the actual condition. The collected data can be analyzed comprehensively with several artificial intelligence algorithms and make data becomes useful information.

Acknowledgements

This research is supported by the collaboration of the computer engineering laboratory and Master of Informatics Islamic State University of Maulana Malik Ibrahim of Malang.

Author details

Fachrul Kurniawan[1]*, Supeno Mardi Susiki Nugroho[2], Mochamad Hariadi[2], Isdaryanto Iskandar[3] and Prita Dewi Basoeki[3]

1 Department of Informatics Engineering, Universitas Islam Negeri Maulana Malik Ibrahim Malang, Indonesia

2 Department of Computer Engineering, Institut Teknologi Sepuluh Nopember, Surabaya

3 Faculty of Engineering, Atma Jaya Catholic University, Jakarta, Indonesia

*Address all correspondence to: fachrulk@ti.uin-malang.ac.id

References

[1] Vu K, Hartley K. Promoting smart cities in developing countries: Policy insights from Vietnam. Telecomm. Policy. 2018;**42**(10):845-859

[2] Susanti R, Soetomo S, Buchori I, Brotosunaryo PM. Smart growth, smart city and density: In search of the appropriate indicator for residential density in Indonesia. Procedia—Social and Behavioral Sciences. 2016;**227**:194-201

[3] Supangkat SH, Arman AA, Nugraha RA, Fatimah YA. The implementation of garuda smart city framework for smart city readiness mapping in Indonesia. Journal of Asia Pacific Studies. 2018;**32**(4):169-176

[4] Kurniawan F, Nugroho SM, Hariadi M. Smart city priority correlation using differential equation. Computing in Science & Engineering. 2018:1-10

[5] Kurniawan F, Nugroho SMS, Hariadi M. Promoting smart city research for engineering students. World Transactions on Engineering and Technology Education. 2019;**17**(1):93-97

[6] Van Der Zee D, Holkenborg B, Robinson S. Conceptual modeling for simulation-based serious gaming. Decision Support Systems. 2012;**54**(1):33-45

[7] Madani K, Pierce TW, Mirchi A. Serious games on environmental management. Sustainable Cities and Society. 2017;**29**:1-11

[8] Berdun FD, Armentano MG. Modeling users' collaborative behavior with a serious game. IEEE Transactions on Games. 2019;**11**(2):121-128

[9] Hu H, Xiao Y, Li H. The effectiveness of a serious game versus online lectures for improving medical students' coronavirus disease 2019 knowledge. Games for Health Journal. 2021

[10] Buijs-Spanjers KR, Hegge HH, Cnossen F, Jaarsma DA, de Rooij SE. Reasons to engage in and learning experiences from different play strategies in a web-based serious game on delirium for medical students: Mixed methods design. JMIR Serious Games. 2020;**8**(3)

[11] Alvarez J, Plantec JY, Vermeulen M, Kolski C. RDU Model dedicated to evaluate needed counsels for Serious Game projects. Computers in Education. 2017;**114**:38-56

[12] Marsh T. Slow serious games , interactions and play: Designing for positive and serious experience and reflection q. Entertainment Computing. 2016:1-9

[13] C. Guestrin, Co-training for semi-supervised learning (cont.) semi-supervised learning. 1-51, 2007

[14] Van De Sande E, Segers E, Verhoeven L. The role of executive control in young children's serious gaming behavior. Computers in Education. 2015;**82**:432-441

[15] Alaribe I. Design a serious game to teach teenagers with intellectual disabilities how to use public transportation. Procedia—Social and Behavioral Sciences. 2015;**176**:840-845

[16] Caragliu A, del Bo C, Nijkamp P. Smart cities in Europe. Journal of Urban Technology. 2011;**18**(2):65-82

[17] Caragliu A, Del Bo CF. Smart innovative cities: The impact of Smart City policies on urban innovation. Technological Forecasting and Social Change. 2019;**142**:373-383

[18] Fernandez-Anez V, Fernández-Güell JM, Giffinger R. Smart city implementation and discourses: An integrated conceptual model. The case of Vienna. Cities. 2018;**78**:4-16

[19] Kang J, Liu M, Qu W. Computers in Human Behavior Using gameplay data to examine learning behavior patterns in a serious game. Computers in Human Behavior. 2017;**72**:757-770

[20] Rüppel U, Schatz K. advanced engineering informatics designing a BIM-based serious game for fire safety evacuation simulations. 2011;**25**:600-611

[21] Wolff A et al. Engaging with the smart city through urban data games. In: Nijholt A, editor. Playable Cities. Gaming Media and Social Effects. Singapore: Springer; 2017. pp. 47-66

[22] García O, Chamoso P, Prieto J, Rodríguez S, de la Prieta F. A serious game to reduce consumption in smart buildings. In: Bajo J, editor. Highlights of Practical Applications of Cyber-Physical Multi-Agent Systems. Berlin: Springer; 2017. pp. 481-493

[23] Viale Pereira G, Cunha MA, Lampoltshammer TJ, Parycek P, Testa MG. Increasing collaboration and participation in smart city governance: a cross-case analysis of smart city initiatives. Information Technology for Development. 2017;**23**(3):526-553

Section 2

Practical Application of Smart Cities Tools and Strategies

Chapter 9

Building Smart System by Applied Deep Learning and Spatial Indoor Agent Based Model for a New Adaptation University Learning Process Post Covid-19

Adipandang Yudono, Sapto Wibowo, Christia Meidiana, Surjono Surjono, Irnia Nurika, Erryana Martati and Yan Akhbar Pamungkas

Abstract

The impact of COVID-19 implied various restrictions on people's mobility, especially for the higher education communities, by implementing the Learning from Home approach. This approach has altered the behavior of a human on a daily basis for a year long. Subsequently, the global vaccination program has been the advent of a "New Normal" approach as it reenables the direct human interactions by following health protocols to abide such as social distancing. This study investigated the pedestrian flow in the Department of Urban and Regional Planning (DURP) lecture building, Brawijaya University, and predicted the potential crowd spots using the Integrated Agent-Based Model (ABM), Computer Vision, and the Geographical Information System on an Indoor scale. Additionally, alternative designs of pedestrian flow were proposed to prevent crowds from occurring. The results showed the East and West entrance paths of the DURP building have high traffic, so the proper response is to organize the Southside door as an alternative entrance for pedestrian access. Moreover, the opening of the south gate could reduce the crowd spots on the 2nd Floor of the DURP lecture building.

Keywords: pedestrian flow, social distancing, new normal, agent-based modeling, computer vision, geographical information system

1. Introduction

The rapid development of science on a global scale has affected the potential evolution of 3 groups of traditional scientific branches formal science, social science, and natural science [1–4]. Nowadays, these scientific branches are associating with each other, thus forming new clusters, such as Social Sciences, Humanities,

Arts for People and The Economic (SHAPE), and the Science-Technology-Engineering-Math (STEM). STEM cluster is formed by the combination of Formal Sciences (Mathematics and Statistics) with Natural Sciences (Biology, Physics, and Chemistry) [5]. In contrast, the SHAPE cluster is formed by the social life linked to scientific fields consisting of politics, psychology, and sociology [6, 7].

The merging of scientific clusters in regards to addressing the global issues related to human life still has some discrepancies. The gaps are still present between the natural science from the STEM cluster and the social science field from the SHAPE cluster. Therefore, the development of a new curriculum consisting of the combination of STEM and SHAPE clusters is proposed, namely Humanitarian Engineering (HE), through the ENHANCE project composed of the works of researchers from the Warwick University of U.K., along with the academics from Greece, Bangladesh, Vietnam, and Indonesia.

In the traditional engineering academic texts, it is challenging to find the term HE thus, some researchers have been trying to define it as the development of science progresses. Passino [8] stated that humanitarian engineering is an approach to constructing technologies that could assist the engineering in helping people, while VanderSteen [9] identified HE as a tool to solve social issues. Moreover, Hill and Miles [10] recognized HE as the solution to social problems by investigating the achievement of sustainability in developing countries. Therefore, this study regards HE as a scientific field that focuses on addressing complex humanitarian matters through the perspective of the SHAPE cluster using a STEM approach to propose smart, equitable, and harmonious solutions.

The purpose of this paper is to analyze the humanitarian engineering field through the micro-scale of the planning field by re-designing the pedestrian flow inside a lecture building concerning the new normal learning process to prevent the higher risk of COVID-19 transmission by avoiding the potential crowd spots using Integrated Agent-Based Model (ABM), Computer Vision, and the Geographical Information System on an Indoor scale.

2. Methods

Replace the entirety of this text with the main body of your chapter. The body is where the author explains experiments, presents and interprets data of one's research. Authors are free to decide how the main body will be structured. However, you are required to have at least one heading. Please ensure that either British or American English is used consistently in your chapter.

This study aims to analyze the lecture building of the Department of Urban and Regional Planning, Brawijaya University, along with its surrounding environment. The descriptive and evaluative analysis is used to investigate the pedestrian's flow, namely Integrated Agent-Based Model (ABM), Computer Vision, and the Geographical Information System on an Indoor scale. The descriptive statistical method is used to investigate the characteristic of the pedestrian consisting of movement, speed, and density. Moreover, evaluative analysis is taken to calculate the density of pedestrian traffic by utilizing the time series data of pedestrians' peak volume during each working hour/day.

Decision making which utilizes images acquired from sensors is known as Computer Vision (CV) [11–13]. The purpose of the CV is to construct an intelligent machine with "see" ability. The Agent-Based Model (ABM) is known as an individual-centric and decentralized approach, whereas the modeler is tasked to pinpoint the agent or active entity (in this case, the person), characterize their behavior, detects agents in a specific environment, create connections in between, and establishes the simulation. Moreover, Geographical Information System (GIS) on an indoor scale is defined as a complete mapping system to make the disconnected project data practical, operate complex artificial environments, track indoor devices, evaluate space allocation in confined spaces, and recognize and react to the real-time events.

The detection of the pedestrian can be achieved by using various computer vision methods. One of the classical human detection methods was invented by the Voila-Jones algorithm, which aims to recognize human faces with a fast detection rate at the cost of low accuracy [14]. It is also revealed that the accuracy drops even more for non-frontal faces [15].

In regards to the development of human detection, the Histogram of Oriented Gradient (GOD) was later proposed in junction with linear Support Vector Machine (SVM), which offers an accuracy rate of up to 89% [16]. In addition, the drawback of the HOG algorithm is required expensive computational resources to operate [17].

Nevertheless, the rapid development of CV human detection through neural network algorithms utilizes the classical methods' base concept, namely, You Only Look Once (YOLO) [18]. Single Shot Detection (SSD) [19], and Faster Region-based Convolutional Neural Network) (Faster R-CNN) [20]. Compared with classical image processing methods, improved robustness and reliability are expected from AI-based human detection [21].

The detection of the pedestrian in this study utilizes the You Only Look Once (YOLO) method. The YOLO algorithm was considered to be used regarding its reliability with a fast detection rate [22]. YOLO has been recognized to be able to outperform HOG-SVM. Therefore, it's been widely used for many purposes, such as operating autonomous cars [23].

YOLO implementation was accessible on DARKNET (open source neural network). The idea of YOLO is to see the whole image once and later passes through the neural network once it immediately detects actual objects, thus later known as the name of the method, YOLO, from the abbreviation of You Only Look Once. The purpose of YOLO is to perform real-time object detection. A localizer or repurpose classifier is used by the detection system. A model is utilized for an image at various locations and scales. The highest-rated area image will be regarded as a detection.

YOLO utilizes the Artificial Neural Network (ANN) approach for object detection in an image by dividing the image into regions and predicting each bounding and probability box from each region. The bounding boxes are later compared with each expected probability. In addition to that, there are several advantages of YOLO compared to a classifier-oriented system. YOLO can carry out the test of the entire image with predictions informed globally on the image. YOLO is also several times faster than the Region Convolutional Neural Network (R-CNN) due to its ability to synthesize the neural network for making predictions than the R-CNN, which needs thousands of images to operate.

In order to represent the actual condition of pedestrian traffic and other geographic features, the ABM is assisted by spatial and geographic visualization data. Several ABM and GIS integration applications have been recognized at the macro scale, such as in the region and urban areas. Hartmann and Zerjav [24] revealed that the assimilation between ABM and GIS is proven effective in planning the optimum health service location concerning the nature of the urban population. The ABM and GIS are generally used for building simulation modeling, such as to estimate the impact of resource investment decisions concerning the health costs, population development of an area, and burden and spread of disease. Nonetheless, the integration of ABM at the microscale with GIS at the indoor scale is still limited. Therefore, to fill this gap, this study aimed to incorporate the main idea of CV as the novelty of the research.

3. Research results

This chapter has three sections: dataset development, human object detection process, and Agent-Based Model and Geographical Information System Indoor. The linkage between these sections will be investigated in the context of research steps for collecting the pedestrian behavior inside the DURP lecture building to rearrange pedestrian traffic to reduce the COVID-19 transmission.

3.1 Dataset development

Datasets collection is the first step before training the YOLO algorithm to recognize the object. The data collection can be conducted through video recording and later exported as images or downloading related pictures from the internet. **Figure 1** describes the collection of the dataset before being used to train the YOLO algorithm.

There are several criteria considered in data processing testing experiments, which consist of:

- Raspberry Pi 3
- Mobile Video Camera
- Video in .mp4 format
- 30 recorded videos condition in total.
- 320 × 240 pixels video resolution with a frame rate of ±20 frames/second.
- Videos were recorded in a stationary state.
- ± 10 seconds of the recoding time of each video.
- There are several conditions to be considered when recording the video, such as indoor environment, dark light (night), outdoors environment or lots of light illumination, and different background conditions.

A camera with a VGA resolution (320 × 240 pixels) and frame rate of ±20 frames/second was used to record the video through the Open Camera application. Static exposure value (camera's sensitivity to light) conditions were used as the camera

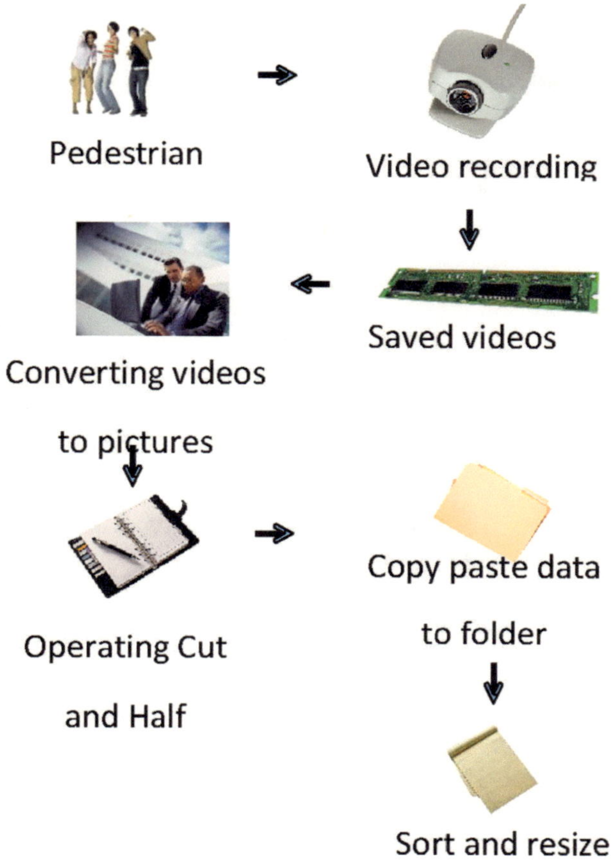

Figure 1.
Video data source capture flow.

settings. The bright condition was recorded during the daytime outside the room and indoors with lights on, while the dark condition was recorded during the evening, and indoor lights were off. The recorded video has a duration of ±10 seconds, and pedestrians walk five meters away from the camera position. The different conditions for recording the video are being considered to investigate the lighting effects, environment (indoor or outdoor), and the presence of other objects than humans, such as shadows, would affect the detection results.

3.2 Human object detection process

The selection of human objects was conducted through blob analysis in MATLAB by examining the size of each object. The purpose of the blob analysis function is to determine the area of a human object based on the minimum and maximum blob area. The elimination of objects with a size of fewer than 10,000 pixels and larger than 980 pixels was considered to eliminate non-human objects. The other stages were conducted, including extraction of video frames, normalization of images, background subtraction, morphological operations, and object detection. Subsequently,

Lightning condition	ΣTest	Detection result	
		ΣTrue	ΣFalse
Bright	10	8	2
Dark	10	2	8
Low	10	1	9
Total	30	11	19

Table 1.
Human object detection testing.

the images extracted from the video frames were normalized. The image was later reconstructed in the form of opening, closing, and filling operations through a background subtraction approach and morphological operations. Specific values are determined at each stage for human object detection.

The conducted test results based on 30 videos were reported in the form of "correct" and "incorrect" information. The test could be considered appropriate if the system detection results match the manual calculation results and vice versa for inappropriate information results. **Table 1** explains the human object detection testing, which revealed 11 incorrect tests out of 30 tests conducted. There were two tests with incorrect information from 10 video tests with bright lighting effects. On the contrary, the number of tests with incorrect information was higher in dark lighting effects, with nine out of 10 tests. These results revealed that the system could detect human (pedestrian) objects better in bright lighting than in darker lighting due to the distraction from the shadows, which were later misinterpreted as human objects. The other distractions that could affect the detection are light bias and other moving objects such as smoke.

The tests during dark conditions revealed that the poor lighting and unstable camera sensor caused the lower detection capability and exceeded the number of detected objects from the manual calculation. Therefore, the DURP lecture building was set with bright conditions (lights on) at the later stage of pedestrian data collection through video recording, as described in **Figure 2**.

3.3 Agen based model and geographical information systems indoor integration process

The agent-based model (ABM) is considered a computational technique that aims to reinforce the analysis of the artificial environment utilized by interfacing the agents in nontrivial ways. The response from every agent is demonstrated separately. Agents who act with agents and later respond to their ongoing case as a set of attitude rules are subsequently derived from the principal theory actions and connections within a definitive framework [25].

When the agent initiates communication with the Geographic Information System (GIS) by sending a "seek to migration" message to the GIS, the evaluation of topological connection and geographic coordinates provides the agent the attitude of being authorized. Subsequently, the GIS responds by updates (renewing the GIS database and related graphical demonstration) or returns a message to the agent mentioning why the migration failed to perform, such as the area is as of now involved, or no permission could be given for the development [26].

DOI: http://dx.doi.org/10.5772/intechopen.106508

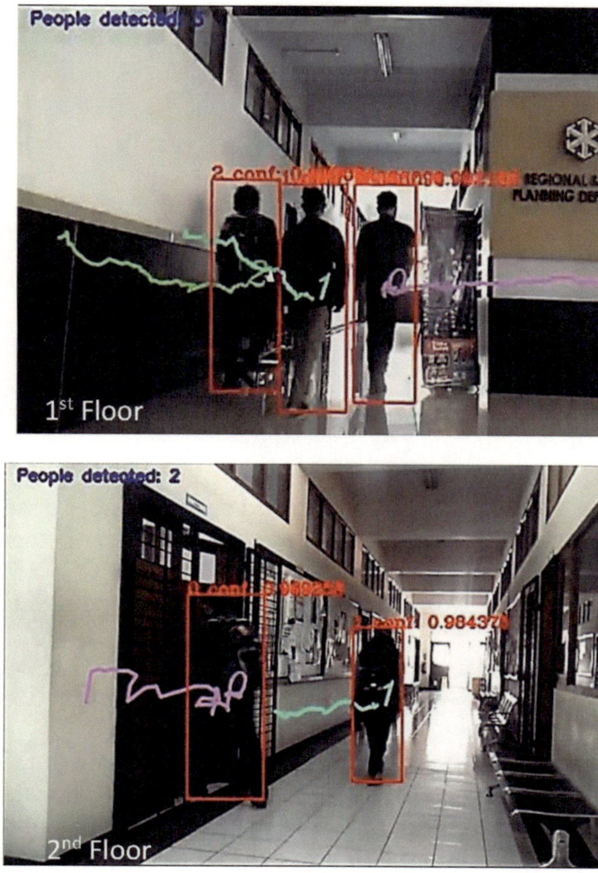

Figure 2.
Video recording of pedestrian detection and tracking on 1st and 2nd floor of the DURP building.

4. Discussions

4.1 Observed pedestrian traffic on the 1st floor of the DURP building under normal conditions

The investigation of pedestrian traffic was done through CCTV recording and later processed with CV to conduct the tracking analysis. It is revealed that the pedestrian traffic on the 1st Floor of the DURP lecture building was concentrated in the corridor connecting the West and East Gate with a very high-density category (1.5). On the contrary, the lowest level of pedestrian density, with values ranging from 0.1 to 0.25, was revealed on the rotation path of the DURP building from the north to south, then turned west towards the Faculty of Engineering Administration building as it is described on **Figure 3**. This route was chosen as the most preferred path by the existing pedestrians because the intersections could connect to many other possible directions.

Another dense pedestrian traffic was recorded in the Plaza area and North-South corridor of the DURP building, representing the traffic from U.B.'s academic community. This was caused by the path connecting the main rooms on the 1st Floor of the DURP building. Considering if the offline teaching and learning process takes place,

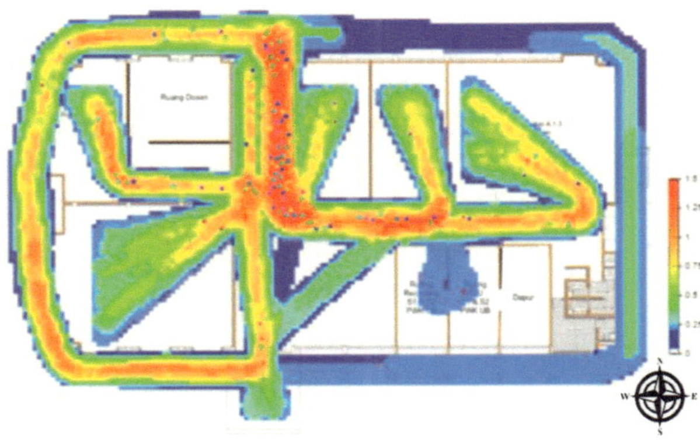

Figure 3.
Observed pedestrian traffic on the 1st Floor of the DURP Building.

then the daily movements could reach 600 based on the number of active students from four grades (2017 to 2020 grade and the average number of students per class is 30 people). Therefore, this situation has the potential for COVID-19 transmission through direct contact, thus worsening the pandemic situation over the university.

4.2 Re-designing pedestrian traffic on the 1st floor of the DURP building for minimizing crowd spots

In response to the observed pedestrian traffic and density in the DURP lecture building, scenarios are proposed to reduce potential physical contact. Opening the south gate with restrictions on the number of the academic community allowed up to 360 from 600 people. Therefore, this proposed scenario could help minimize the COVID-19 virus transmission through potential physical contact.

In addition to the proposed scenario, offline lecture classes are designed for students from grade 1 (batch 2020) and grade 2 (batch 2019), whereas grades 3 and 4 (batch 2018 and 2017) are proposed for online learning procedures. Finally, the simulation of integration of ABM and GIS Indoor scale to re-design pedestrian traffic on the 1st floor could significantly decrease pedestrian traffic and density, as explained in **Figure 4**.

4.3 Observed pedestrian traffic on the 2nd floor of the DURP building under normal conditions

Based on the observed pedestrian traffic on the 2nd floor of the DURP building, the stairway and plaza in front of the stairs and the route to the DURP library are to be considered to be traversed, as explained in **Figure 5**.

It is revealed that the plaza in front of the stairs, the stairway, and the DURP library are the most preferred route by pedestrians because there are no classrooms available on the 2nd floor, only consists of lecturer's room, library, and the academic's meeting rooms. Therefore, a low magnitude of pedestrian traffic was observed through the North-South corridor area. Furthermore, a high magnitude of pedestrian traffic is expected on the 3rd floor since it primarily consists of classrooms and a computer laboratory.

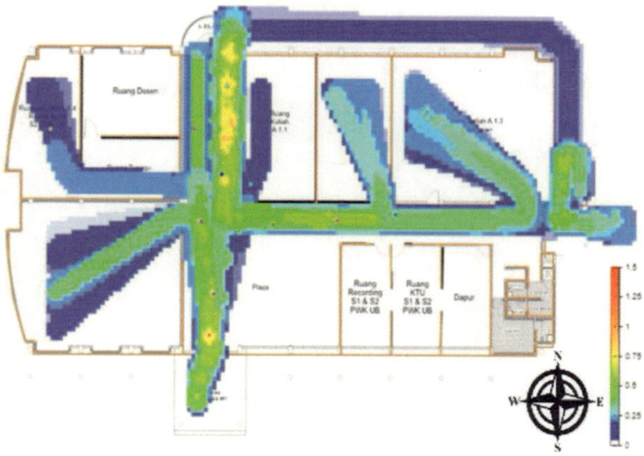

Figure 4.
Re-designing the access for minimizing pedestrian traffic on the 1st floor of the DURP building.

Figure 5.
Observed pedestrian traffic on the 2nd floor of the DURP building.

4.4 Re-designing pedestrian traffic on the 2nd floor of the durp building for minimizing crowd spots

According to the observed pedestrian traffic on the 2nd floor, scenarios are proposed to re-design the traffic. The restriction of the number of the academic community and the opening of the east gate could produce a significant decrease of pedestrian traffic, especially in the plaza on the north and the staircase corridor, as explained in **Figure 6**.

The proposed scenario could help reduce the pedestrian traffic in the staircase corridor at the north, since there is no classroom on the 2nd floor of the DURP building. Thus, dividing the entrance to the 2nd floor, which formerly could only be accessed through the staircase, could help reduce the pedestrian traffic and later could help on minimizing the potential transmission of COVID-19.

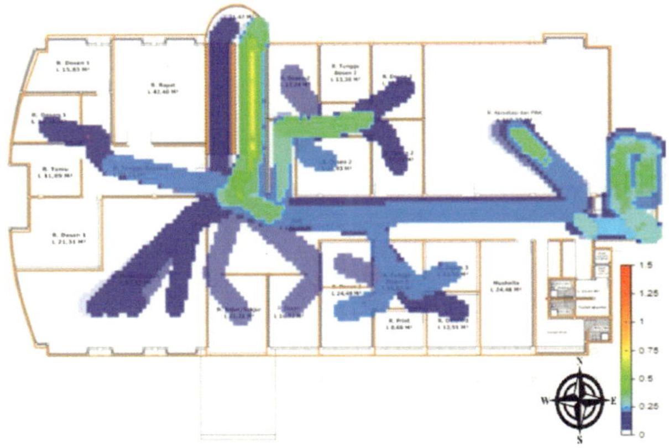

Figure 6.
Re-designing the access to minimize pedestrian traffic on the 2nd floor of the DURP building.

5. Conclusion

The relatively new scientific field of Humanitarian Engineering (HE) could potentially fill the gaps between STEM and SHAPE clusters. HE aims to investigate humanitarian trends and issues mainly studied in the social science field from the SHAPE cluster and later proposed the solution through engineering perspective from natural science and formal science from the STEM cluster. HE approaches are considered in this paper, referring to the latest social issues concerning the COVID-19 pandemic. Human activities such as school shopping and work are expected to return to normal once the global vaccination program has been completed. On the contrary, there were no optimal results reported from the studies related to the impact of global vaccination. Therefore the 'new normal' approach is proposed while maintaining the health protocols, in which avoiding crowd spots is part of the protocols.

It is revealed that the HE approach by studying pedestrian traffic in the DURP lecture building through CV and ABM-GIS Indoor simulation could helps on minimizing the crowd spots. The north and west entrance paths on the 1st Floor were observed with a high magnitude of pedestrian traffic. Therefore, the east side door opening could be an alternative for new accessibility for pedestrians. A similar approach applies to the 2nd Floor by opening the east gate could help minimize crowd spots.

The restrictions on the number of academic communities entering the DURP building are considering the need for empirical and site visits to case studies for grades 1 and 2. In addition, grades 3 and 4 have fewer classes to attend, and the learning patterns of senior students are emphasized critical thinking through the exploration of literature studies outside the classes.

Acknowledgements

This study was conducted by developing Humanitarian Engineering Curriculum under ENHANCE Project, funded by Erasmus+ with partners from Warwick University (U.K.), University of West Attica (Greece), Universitas Brawijaya (Indonesia), Institut Teknologi Bandung (Indonesia), Universitas Gadjah Mada (Indonesia), Bangladesh University of Engineering and Technology (Bangladesh),

University of Dhaka (Bangladesh), Ho Chi Minh City University of Transport (Vietnam), and Ho Chi Minh City University of Technology (Vietnam). Furthermore, this research combined with Universitas Brawijaya's research under the Faculty of Engineering, Universitas Brawijaya's non-tax revenue (PNPB) research scheme. Therefore, at the end of this writing, the researchers would thank Erasmus+ and the Faculty of Engineering—Universitas Brawijaya.

Author details

Adipandang Yudono[1,2*], Sapto Wibowo[3], Christia Meidiana[1], Surjono Surjono[1], Irnia Nurika[4], Erryana Martati[5] and Yan Akhbar Pamungkas[6]

1 Department of Urban and Regional Planning, Brawijaya University, Indonesia

2 Center of Geospatial Information Systems and Data Science, Brawijaya University, Indonesia

3 Department of Electronic Engineering, State Polytechnic of Malang, Indonesia

4 Department of Agro Industrial Technology, Brawijaya University, Indonesia

5 Department of Agricultural Product Technology, Brawijaya University, Indonesia

6 Center of Geospatial Information Systems and Data Science, Brawijaya University, Indonesia

*Address all correspondence to: adipandang@ub.ac.id

References

[1] van de Lagemaat R. Theory of Knowledge for the I.B. Diploma. Cambridge: Cambridge University Press; 2006. ISBN 978-0-521-54298-2

[2] Popper KR. The Logic of Scientific Discovery. New York, NY: Routledge Classics; 1959. ISBN 0-415-27844-9. OCLC 59377149

[3] Castelvecchi D. Nature magazine (2015-12-23). "Is string theory science?". Scientific American. 2015. Available from: https://www.scientificamerican.com/article/is-string-theory-science/ [Retrieved: 31st May 2021]

[4] Cohen E. The boundary lens: Theorising academic activity. In: The University and its Boundaries: Thriving or Serving in the 21st Century. 1st ed. New York: Routledge; 2021. pp. 14-41. ISBN 0367562987

[5] Hallinen J (n.d.). STEM education. Encyclopedia Britannica. Available from: https://www.britannica.com/topic/STEM-education [Retrieved: 31st May 2021]

[6] British Academy. "SHAPE". SHAPE; 2020. Retrieved: 31st May 2021

[7] Black J. SHAPE—A focus on the human world. Social Science Space. 2020. Available from: https://www.socialsciencespace.com/2020/11/shape-a-focus-on-the-human-world/ [Retrieved: 31st May 2021]

[8] Passino KM. Humanitarian Engineering: Creating Technologies that Help People. 2nd ed. Columbus, Ohio: Bede Publishing; 2015

[9] VandeSteen JDJ. Humanitarian Engineering in the Engineering Curriculum. Doctoral of Philosophy Thesis at Queen's University, Kingston, Ontario, Canada. August 2008

[10] Hill S, Miles E. What do students understand by the term 'Humanitarian Engineering'?. EE2012. International Conference on Innovation, Practice and Research in Engineering Education; 2012

[11] Shapiro LG, Stockman GC. Computer Vision. New York, USA: Prentice-Hall, Inc.; 2001

[12] Forsyth DA, Ponce J. Computer Vision: A Modern Approach. 2nd ed. Boston, USA: Pearson Education Ltd.; 2011

[13] Szeliski R. Computer Vision: Algorithms and Applications. Washington, USA: Springer; 2011

[14] Viola P, Jones M. Robust Real-Time Object Detection. Second International Workshop of Statistical and Computational Theories of Vision–Modeling, Learning, Computing, and Sampling, Voncouver, Canada. July 13, 2001

[15] Shan T, Lovell BC, Chen S. Face recognition robust to head pose from one sample image. In: 18th International Conference on Pattern Recognition. 2006. pp. 515-518. DOI: 10.1109/ICPR.2006.527

[16] Dalal N, Triggs B. Histograms of oriented gradients for human detection. In: 2005 IEEE Computer Society Conference on Computer Vision and Pattern Recognition. Vol. 1. 2005. pp. 886-893. DOI: 10.1109/CVPR.2005.177

[17] Rettkowski J, Boutros A, Göhringer D. HW/sw co-design of the hog algorithm on a xilinx zynq soc. Journal of Parallel and Distributed

Computing. 2017:109. DOI: 10.1016/j.jpdc.2017.05.005

[18] Redmon J, Divvala S, Girshick R, Farhadi A. You only look once: Unified, real-time object detection. In: 2016 IEEE Conference on Computer Vision and Pattern Recognition. 2016. pp. 779-788. DOI: 10.1109/CVPR.2016.91

[19] Liu W, Anguelov D, Erhan D, et al. SSD: Single shot multibox detector. In: Leibe B, Matas J, Sebe N, Welling M, editors. Computer Vision-ECCV 2016. Lecture Notes in Computer Science. Cham: Springer; October 8-16, 2016. pp. 21-37. DOI: 10.1007/978-3-319-46448-0_2

[20] Ren S, He K, Girshick RB, Sun J. Faster r-cnn: Towards real-time object detection with region proposal networks. In: Advances in Neural Information Processing Systems. 2015. pp. 91-99

[21] Uçar A, Demir Y, Güzeliş C. Object recognition and detection with deep learning for autonomous driving applications. Simulation. 2017;**93**(9):759-769. DOI: 10.1177/0037549717709932

[22] Srivastava S, Divekar AV, Anilkumar C, Naik I, Kulkarni V, Pattabiraman V. Comparative analysis of deep learning image detection algorithms. Journal of Big Data. 2021;**8**(1):66. DOI: 10.1186/s40537-021-00434-w

[23] Al-Nuaimi M, Wibowo S, Qu H, Aitken J, Veres S. Hybrid verification technique for decision-making of self-driving vehicles. Journal of Sensor and Actuator Networks. 2021;**10**(3). DOI: 10.3390/jsan10030042

[24] Hartmann T, dan Zerjav V. Optimizing the location of out-care centers in urban space using agent-based modeling. In: Construction Research Congress 2014 ©ASCE. 2014. pp. 2375-2384

[25] Macal CM, North MJ. Tutorial on agent-based modelling and simulation. In: Euhl ME, Steiger NM, Armstrong FB, Joines JA, editors. Proceedings of the 2005 Winter Simulation Conference. 2005

[26] Brown DG, Riolo R, Robinson DT, North M, Rand W. Spatial process and data models: Toward integration of agent-based models and GIS. Journal of Geographical Systems, Special Issue on Space-Time Information Systems. 2005;**7**(1):25-47

Chapter 10

The Role of Aggregators in Smart Grids

Lasse Berntzen and Qian Meng

Abstract

Sustainable smart cities need to focus on energy production and use. By installing solar panels, prosumers may contribute to the energy production in the city. The use of solar panels is particularly relevant to free-standing residential buildings. Prosumers may also trade flexibility, the ability to shift energy use to periods when the total energy consumption is lower. Prosumers may also store energy for future sale or consumption. An aggregator is a new role connecting prosumers with energy providers. The aggregator negotiates terms, provides flexibility on behalf of its prosumers, and may even provide energy storage capabilities. This chapter describes the evolving role of aggregators and their possible business models. The aggregators will contribute to smarter energy production and use in smart cities.

Keywords: aggregator, flexibility, prosumer, smart cities, smart energy, smart grids, sustainability

1. Introduction

In the ERA-NET Smart Multi-layer Aggregator project [1], the University of South-Eastern Norway (USN) was leading a work package on emerging business models in the energy sector [2] and also performed research on the adoption of new technology to achieve more flexibility within electricity grids [3]. Before this project, we had been researching electronic government, refocusing on smart cities since 2015 [4]. During the ERA-NET project, we soon saw the emerging role of the aggregator as a new actor in the energy market. Smart energy is closely connected to smart cities. Improving energy efficiency is one of the obvious ways of being smart. Smart energy has become more important with energy shortage and increasing energy prices.

The utilization of energy in a city is a complex process. A modern city needs to fulfill the demand for energy for purposes of commerce, household, infrastructure, transport, etc. Sustainable energy, especially solar energy produced in households and other buildings, has changed the current energy market and plays a significant role in the energy landscape for a smart city [5]. Solar power is primarily used for electric energy generation, but a small fraction of the solar power is used for thermal energy.

Statistics published by the International Renewable Energy Agency (IRENA) [6] show that in the last ten years, worldwide solar energy generation capacity increased with a steady high annual growth both for solar photovoltaic and solar thermal; even in the year 2020 when COVID-19 struck the world heavily. The change is shown

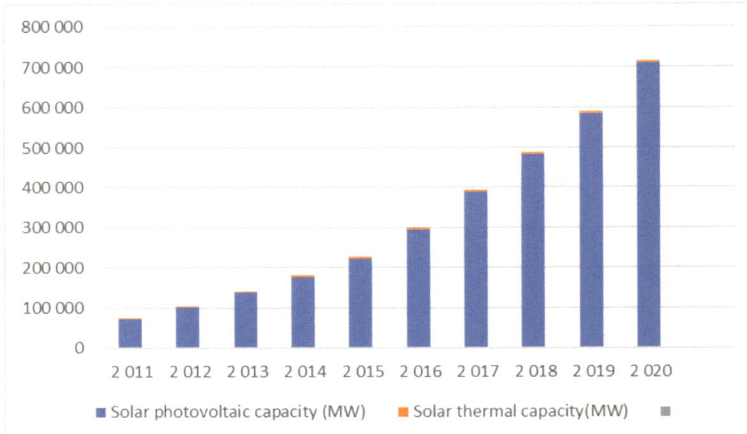

Figure 1.
Global growth in solar energy capacity.

in **Figure 1**. Solar photovoltaic energy dominates renewable capacity expansion, accounting for around 100 GW installation capacity growth in 2018–2020.

The next section discusses smart energy as one application domain of smart cities. The third section discusses smart grids and the emerging role of prosumers. The fourth section focuses on the role of aggregators. Section five brings in the topic of electric vehicles in smart cities, and finally, the sixth section provides a conclusion.

2. Smart cities and smart energy

The smart city is a concept with no unified definition [7], but technology plays an important role. The main objective is to improve the quality of life for its citizens through better service provision, reduced environmental footprint, and improved participation. In most cases, smart cities are materialized through projects within application areas [8]. Such application areas can be smart traffic [9], smart parking [10], smart public transport [11], smart waste handling [12], smart safety and security [13], and smart energy, as shown in **Figure 2**.

Smart energy is a concept where information and communication technologies are used to achieve the process of using devices for energy efficiency. Smart energy is about reducing energy use but also introducing new renewable energy sources.

Smart energy relies on smart grids to improve energy efficiency mainly by adopting smart meters (SM) that allow almost real-time tracking of power consumption. SM can also monitor and control the electric power consumption of appliances. In addition, SM can measure power production from solar panels and power transmission from electric vehicles. During the shift of grids toward smart grids, SM enable customers to use electric power more efficiently, but also contribute to energy pricing based on current demand in the market. This will incentivize customers to plan their consumption and thereby contribute to increased flexibility.

Calvillo, Sánchez, and Villar [14] propose a comprehensive smart city model that includes all energy-related activities while keeping the size and complexity of the model manageable. Such a model is highly desirable to successfully meet the

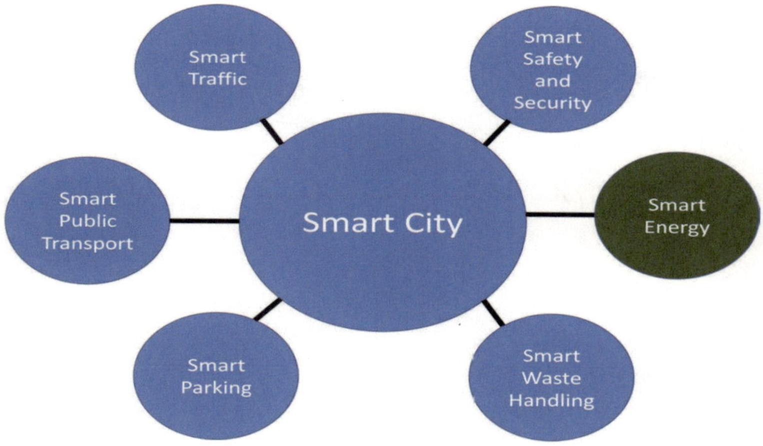

Figure 2.
The smart city and some application domains.

increasing energy needs of present and future cities. They propose five main energy-related activities that have been called intervention areas:

- Generation
- Storage
- Infrastructure
- Facilities
- Transport (mobility)

All these areas are related to each other but contribute to the energy system in different ways: generation provides energy, while storage helps in securing its availability; infrastructure involves the distribution of energy and user interfaces; facilities and transport are the main final consumers of energy, as they need it to operate.

3. Smart grids and prosumers

The energy distribution normally follows a two or three-layer model, where the top level is the transmission system operator (TSO). In Norway, Statnett is the transmission system operator. Statnett is a state enterprise owned by the Norwegian state through the Ministry of Petroleum and Energy. The mission of Statnett is to secure the Norwegian power supply through operations, monitoring, and preparedness. Statnett also plays an essential role in realizing Norway's climate objectives [15].

The distribution system operators connect the customers to the grid. In 2018 Norway had 124 distribution system operators, with the ten largest having two-thirds of the customers and 60% of the total energy deliveries [16]. **Figure 3** shows the organization of the electricity grid.

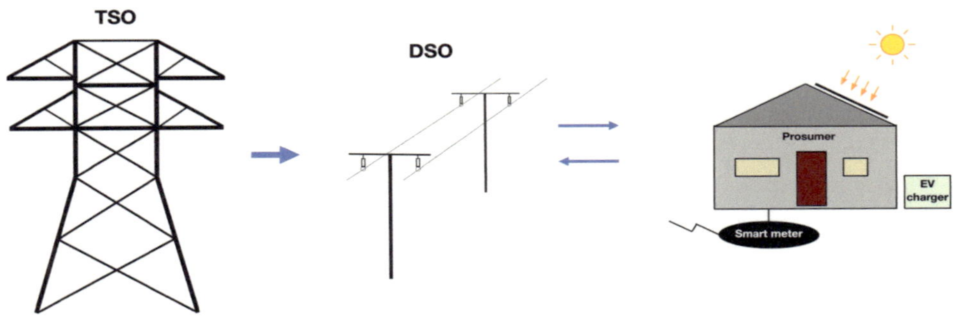

Figure 3.
Electricity flow in the smart grid.

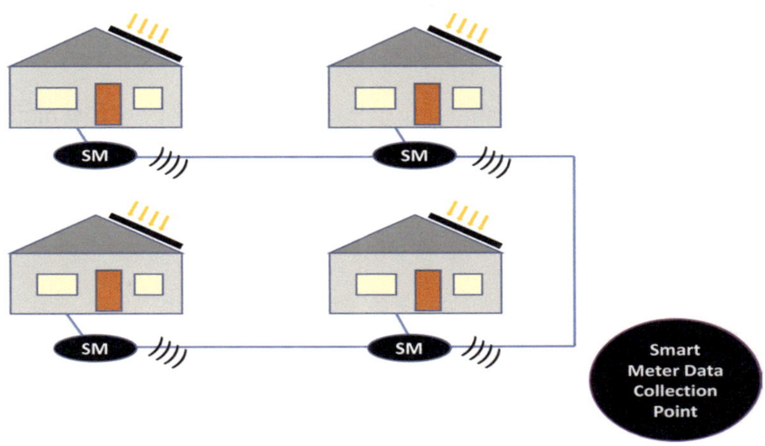

Figure 4.
Smart grid.

A smart grid is a local electricity grid enabling a two-way flow of electricity and data, including various operation and energy measures, such as SM, smart appliances, renewable energy resources, and energy-efficient resources [17].

Figure 4 shows a smart grid consisting of four households with installed photovoltaic panels and SM. The SM communicate with a smart meter data collection point through wireless technology or power line communication. The households are examples of prosumers since they can both produce and consume electric energy.

The most common production comes from solar panels, but prosumers can also generate electricity from wind and geothermal wells. Renewable energy is dependent upon environmental conditions. These conditions vary with the hours of the day and the weather.

In the context of smart cities, prosumers are not restricted to households. All buildings, including apartment blocks, office buildings, and shopping malls, can be prosumers. As long as they have open areas exposed to the sun, they may produce energy. Also, geothermal energy may be an option since geothermal energy can be used both for heating and cooling purposes.

Flexibility is when a building or household can change consumption patterns based on the situation in the energy market. Flexibility can be precious to the DSO to handle possible peaks.

Load shifting and peak shaving are two important techniques to improve energy use [18]. The consumption of electricity varies throughout the day. In Norway, we have one peak in the morning and one in the afternoon. The peak in the morning is mainly caused by electric water heaters kicking in after morning showers. The afternoon peak happens when most are coming home from work, cooking dinner, etc. Load shifting can simply be explained as moving the load to other time periods. One example is to spread the load from water heaters. Another example is to charge electric vehicles during the night.

Figure 5 shows load shifting in practice. The goal is to keep the consumption at a maximum of 5.5 units throughout the day. The morning peak, from 07:00 to 10:00, is above this level. By controlling water heating, some of the load can be shifted to later in the day. A total of 8.5 units need to be moved. From 12:00 until 16:00, the consumption is lower than 5.5 units, so the spare capacity can be used for water heaters and other appliances that have been put into flexibility mode. Note that the limit of 5.5 units is not fully utilized from 13:00 to 16:00. A new peak appears from 17:00 to 19:00. For this period, a total of 3.5 units need to be shifted to later in the evening,

Peak shaving has to do with storage. If electricity can be stored somewhere in the facilities, this stored energy can be used to shave the peaks. Tesla has introduced its power wall as a household battery storage system [19]. Storage can also be centralized, as described in the next section. Load shifting and peak shaving are important since the grid needs to be dimensioned to handle the peaks.

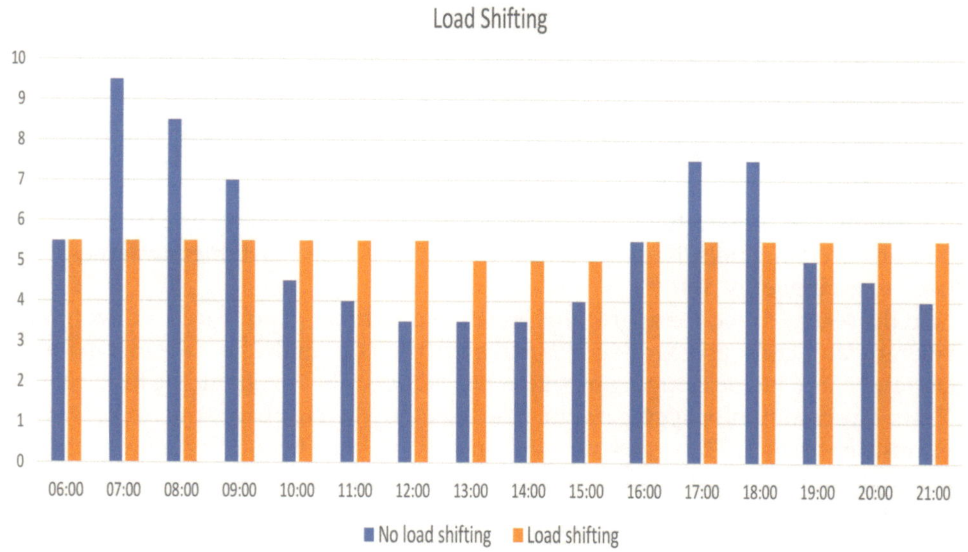

Figure 5.
Load shifting.

4. The role of aggregators

A prosumer can produce and sell surplus energy in the energy market. In Norway, a prosumer can sign an agreement (energy customer plus agreement) with an energy provider. The distribution system operator is obliged to facilitate energy transfer to and from the customer. The prosumer cannot produce more than 100 kWh per hour. If production exceeds this limit, the prosumer must seek a license as an energy producer [20]. The tariffs for selling energy to the market are generally not beneficial. A prosumer will seek to use its own produced energy before selling to the market.

To facilitate collaboration among a group of prosumers, an aggregator is necessary. The EU 2019/944 Electricity Directive defines aggregation as a "function performed by a natural or legal person who combines multiple customer loads or generated electricity for sale, purchase, or auction in any electricity market" [21].

The role of aggregators and their function has become a hot topic for the reason of being a significant part of the European power market. The European framework assigns aggregators a fundamental role in energy market liberalization and distributed energy resources (DER) integration toward carbon-neutral energy systems. The aggregators cannot only participate in the demand response activity and wholesale market bidding but also contribute to maximizing economic efficiency and fostering cross-zonal trading, considering, in particular, the overall system efficiency [22].

As the number of prosumers grows, the business opportunities for a new energy ecosystem actor, the aggregator, emerge. As earlier mentioned, flexibility may be important to shift or shave peaks caused by differences in consumption during the day. The aggregator is a business entity that can aggregate energy from a group of prosumers. A higher volume benefits the aggregator when negotiating with the distribution system operators and energy providers. The aggregator can also provide services, such as settlements, storage, etc.

4.1 Smart-MLA

The Smart-MLA project [1] made a prototype for a multi-layer aggregator, as shown in **Figure 6**.

The main goal of the ERA-NET Smart Multi-Layer Aggregator project (Smart-MLA) was to demonstrate how an aggregator could improve energy efficiency through flexibility [2]. On the lowest layer, the community aggregator simply collects information from a household to optimize energy use. The primary impact is the reduction of energy bills, but a secondary effect is increased customer awareness related to their energy use.

On the second layer, the aggregator will start controlling appliances. The community aggregator collects usage patterns and constraints from the household. One such constraint could be the charging of an electric vehicle. The car should be fully charged at 7 am. The community aggregator can then decide when to charge the car as long as the constraint is met. The aggregator will consider the future hour-by-hour energy prices and information about the weather to predict output from photovoltaic panels.

On the third layer, the aggregator does not only optimize the flexibility obtained by controlling appliances but also uses this flexibility to negotiate with the market. The aggregator uses the combined flexibility of its prosumers to improve the market position. The aggregator handles settlements with its prosumers and the energy provider.

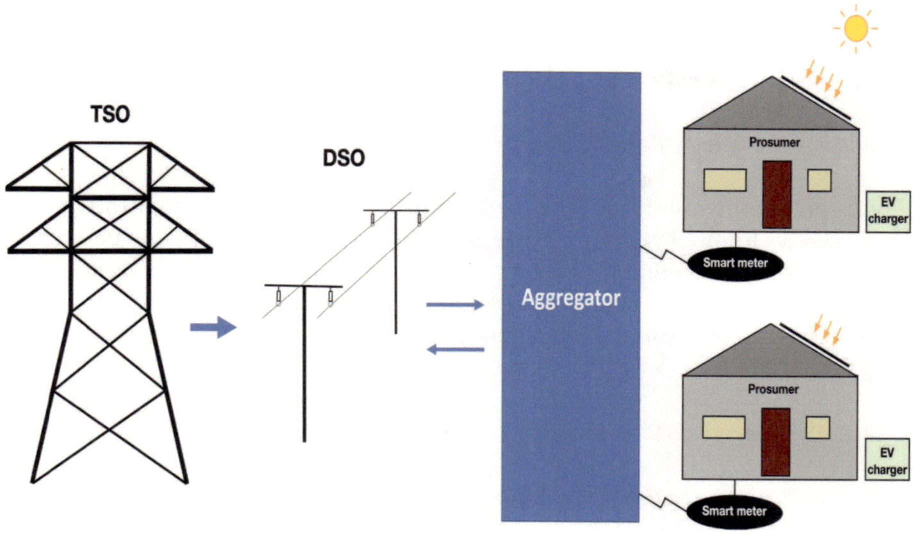

Figure 6.
The aggregator in the smart grid.

While the project demonstrated the opportunities of the multi-layer aggregator model, there are still barriers to overcome. First, there are regulatory issues that need to be handled. The aggregator needs to be established as part of the energy ecosystem. Using Norway as a selected case, the current regulatory environment does not recognize the aggregator as a market actor. A prosumer can sell energy to an energy provider within the limit of 100 MWh per hour. An aggregator will easily exceed this limit and has to be licensed as an energy provider.

The second obstacle is the lack of trust in the energy market. As part of the project, the University of South-Eastern Norway surveyed early adopters of smart home technology [3]. The results showed that the early adopters wanted full control of their energy production and consumption and were unlikely to transfer control to an external entity. Based on the results, we discussed possible remedies from organizational models where the prosumers own the aggregator as a cooperative. Regulatory measures include self-regulation to make the energy market more transparent to achieve the necessary trust among the prosumers. Also, technology can be used to increase trust. Our research also pointed to blockchain technology as a possibility to achieve full transparency about pricing and settlements.

4.2 Aggregator business opportunities

The main electricity market stakeholders include the power generators, transmission system operators, distribution system operators, prosumers, and aggregators. In the transition to green energy, there are great business opportunities for aggregators, which can be categorized as follows [2]:

4.2.1 Energy efficiency services provider

The aggregator offers the customers an energy-saving plan by installing high-efficiency equipment. The aggregator can monitor and control the equipment to participate in the demand response in the power market. E.g., at the peak of the power consumption, the aggregator helps users to reduce their demand and consume electricity later when the power price is low.

4.2.2 Information value-added services provider

The aggregator provides their consumers a value-added service through IoT and big data technologies that provide data and analysis for real-time electricity prices, electricity demand and consumption at a household, and power distributed generation nearby. Then the prosumers can take control of their electricity consumption in real-time and decide when to sell their own generated power at a peak in the grid.

4.2.3 Integrated energy services provider

With the demand for green transition and access to various smart terminals like electric vehicles, charging stations, smart home appliances, and distributed energy generation, the aggregator can develop the business to cooperate with other service suppliers (like heating) to deliver integrated energy services, optimize the integrated energy solution to maximize the benefit for the users. With many assets and a wide range of businesses, the aggregator may behave in the dual role of an energy supplier and an energy service provider.

4.2.4 Extended services provider for zero-emission

For the EU zero-emission target, many countries have implemented policies and measures to replace fossil fuel cars with electric cars. In practice, Norway's electric vehicle policy has proven effective by reducing taxes and fees for electric vehicles while fossil fuel cars are heavily taxed. Thus electric vehicles have become much cheaper than fossil fuel cars. As a result, by the end of 2021, there were 460,734 electric cars registered among a total of 2,893,987 private passenger cars [23]. This clearly shows how incentives shape consumer choice by a combination of taxes and rewards. The aggregator can then expand their customer channels through cooperation with the electrical vehicle sellers and benefit prosumers with their energy-saving, information, and integrated energy services.

4.3 The aggregator as a storage provider

In the smart grid, the intermittent and random output of solar energy has brought challenges to the balance of demand, supply, and grid stability. As to prosumers, solar energy is stored for self-consumption in most cases. While from the perspective of energy efficiency and management efficiency, storing energy by the aggregator will be a more feasible solution [24], as shown in **Figure** 7.

In storage service, prosumers store energy mostly for self-consumption. Even if they make a profit out of the outrage of storing produced energy in the battery and selling energy at peak time to maximize their own profit, this could be inefficient when taking many prosumers as a system. Scale effect also works with aggregation of many prosumers than respectively. For prosumers, it is not only the cost of batteries but also the additional hardware to handle the charging and discharging of the batteries and the installation cost that need to be considered when investing in battery

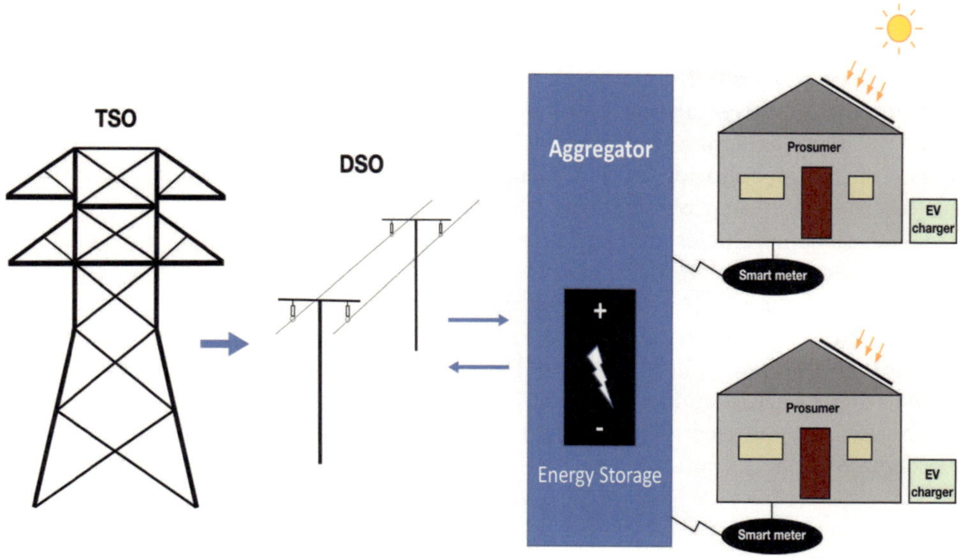

Figure 7.
Energy storage by the aggregator.

storage. If an aggregator supplies a storage service, the aggregator could use a larger facility and not be overly concerned about the compactness of the installation [25].

In addition to achieving the outrage goal, aggregators storing electricity is also a key mechanism for supplying electricity reliably, increasing security and economic value, and decreasing carbon dioxide emissions. Aggregator storage also plays a significant role in keeping a balance between supply and demand, avoiding electric fluctuations, contributing to the stability of the low voltage DSO grid, and making the DSO grid system more efficient, especially for the weak low voltage grid in Norway [24].

5. Electric Vehicles in the Smart City – Norway as a Case Study

Electro-mobility is an important exponent of smart city strategies. Considerable investments in electric vehicles are being made worldwide, and supporting infrastructure not only offers the potential to reduce road transport emissions but also unlock other smart city opportunities. This includes new solutions for mobility, energy use, public services, residential and commercial buildings, wider urban systems, citizen engagement, and behavior change. Accelerating the adoption of electric vehicles, and realizing the associated smart city benefits, requires coordinated action among all stakeholders [26].

One case for smart energy use in smart cities is the adoption of electric vehicles. Electric vehicles reduce the environmental footprint by reducing CO2 emissions and other air pollutants. Norway has the highest adoption rate of electric vehicles worldwide. This result is due to the Norwegian government's determination and effective measures.

The Norwegian Parliament has decided on a national goal that all new cars sold by 2025 should be zero-emission (electric or hydrogen). By February 2022, there were more than 470.000 registered battery electric cars (BEVs) in Norway. Battery electric vehicles held a 64 % market share in 2021. The transition speed is closely related to policy instruments and a wide range of incentives [27].

Five years ago, Oslo, the Norwegian capital, had some serious problems with air pollution caused by certain meteorological conditions. In January 2017, Oslo was closed for diesel cars for a short period. The city council considered raising traffic tolls on days with high pollution levels. The uptake of electric vehicles has significantly reduced the pollution problems seen earlier.

When electric vehicles are considered to contribute to smart cities for energy storage and green transition. Tesla Powerwall is the pioneer with its battery based on lithium iron phosphate (LiFePO4) chemistry. With the development of green energy, battery technology is also undergoing a significant transformation. According to BloombergNEF's research, lithium-ion battery pack prices were above $1,200 per kilowatt-hour in 2010 and fell 89% to $132/kWh in 2021 [28].

Therefore, soon, electric vehicles are expected to meet the EU zero-emissions goal, serving as one part of smart energy and a role of energy storage in the smart grid.

6. Conclusion

Smart energy is an important part of smart cities. Smart cities need to be energy efficient. The role of prosumers refers to buildings and households that can produce renewable energy. The aggregator is a new role in the energy ecosystem. The aggregator can represent a group of prosumers dealing with the energy market. The aggregator may also offer additional services to help its prosumers achieve more energy efficiency. While fulfilling the balance between the energy demand and supply, especially for load shifting and peak shaving, energy storage is an important component. Prosumer storage is efficient for self-consumption mode, but from the perspective of scale effect for many prosumers, storage provided by an aggregator is more feasible and sustainable. New business opportunities for the aggregators have been identified, and aggregators will play a significant role under the EU framework to achieve the green transition goals. Electric vehicles will also contribute to smart traffic and smart energy when their worldwide adoption increases.

The function and role of aggregators in a smart city need more investigation, such as the social acceptance of aggregators in the energy market, the interaction and collaboration with other stakeholders, and creating business models for aggregators. The fundamental role of aggregators in the European power market and distributed energy resources will become clearer.

Acknowledgements

This work was supported by the Manu Net scheme Grant number MNET20/NMCS-3779 and funded through the Research Council of Norway Grant number 322500 with the project title "Cloud-based analysis and diagnosis platform for photovoltaic (PV) prosumers." It builds on results from the ERA-Net Smart Grids Plus scheme Grant number 89029 with the project title "Multi-layer aggregator solutions to facilitate optimum demand response and grid flexibility," funded through the Research Council of Norway Grant number 295750.

Author details

Lasse Berntzen* and Qian Meng
University of South-Eastern Norway, Horten, Norway

*Address all correspondence to: lasse.berntzen@usn.no

References

[1] Smart-MLA Consortium. SMART-MLA Multi-Layer Aggregator Solutions to Facilitate Optimum Demand Response and Grid Flexibility. 2022. Available from: https://eranet-smartenergysystems.eu/global/images/cms/Content/Fact%20Sheets/ERANetSES_SMART_MLA.pdf. [Accessed: July 13, 2022]

[2] Berntzen L, Meng Q, Johannessen MR, Vesin B, Brekke T, Laur I. Aggregators and prosumers – An analysis of business model opportunities. In: Proceedings of the 2021 International Conference on Engineering and Emerging Technologies (ICEET). 2021. pp. 1-6

[3] Johannessen MR, Berntzen L, Vesin B, Meng Q, Brekke T, Laur I. User Sentiments Towards Smart Grid Flexibility – A survey of early adopters' attitude towards allowing third parties to control electricity use in households. In: Proceedings of the Fourteenth International Conference on Advances in Human-oriented and Personalized Mechanisms, Technologies, and Services (CENTRIC). 2021. pp. 41-46

[4] Berntzen L, Johannesen MR. The role of citizen participation in municipal smart city projects: Lessons learned from Norway. In: Gil-Garzia R, Pardo T, Nam T, editors. Smarter as the New Urban Agenda: A Comprehensive View of the 21st Century. Cham, Switzerland: Springer; 2015. pp. 299-314

[5] Axon CJ, Roberts SH. Energy for cities: Supply, demand and infrastructure investment. In: Bishop J, editor. Building Sustainable Cities of the Future. Green Energy and Technology. Cham, Switzerland: Springer; 2017. pp. 5-27

[6] International Renewable Energy Agency (IRENA). Renewable Capacity Statistics 2022. 2022. Available from: https://irena.org/publications/2022/Apr/Renewable-Capacity-Statistics-2022. [Accessed: July 13, 2022]

[7] Kobashi T et al. Smart city and ICT infrastructure with vehicle to X applications toward urban decarbonization. In: Yamagata Y, Yang P, editors. Urban Systems Design: Creating Sustainable Smart Cities in the Internet of Things Era. Amsterdam, Netherlands: Elsevier; 2020. pp. 289-334

[8] Johannessen MR, Berntzen L, El-Gazar R. A survey on smart cities, big data, analytics, and smart decision-making – Towards an analytical framework for decision-making in smart cities. International Journal on Advances in Intelligent Systems. 2019;**12**(1-2):27-38

[9] Flore U, Florea A, Lechuga GP. An interdisciplinary review of smart vehicular traffic and its applications and challenges. Journal of Sensor and Actuator Networks. 2019;**8**(1):13

[10] Diaz Ogás MB, Fabregat R, Aciar S. Survey of smart parking systems. Applied Sciences. 2020;**10**:3872

[11] Kuo Y-H, Leung JMY, Yan Y. Public transport for smart cities: Recent innovations and future challenges. European Journal of Operational Research. 2022 [Online]

[12] Fallavi KN, Kumar RV, Chaithra BM. Smart waste management using Internet of Things: A survey. In: Proceedings of the 2017 International Conference on I-SMAC (IoT in Social, Mobile, Analytics and Cloud) (I-SMAC). 2017. pp. 60-64

[13] Lacinák M, Ristvej J. Smart city, safety and security. Procedia Engineering. 2017;**192**:522-527

[14] Calvillo CF, Sánchez-Miralles A, Villar J. Energy management and planning in smart cities. Renewable and Sustainable Energy Reviews. 2016;**55**:273-287

[15] Statnett. About Statnett. Available from: https://www.statnett.no/en/about-statnett/. [Accessed: July 13, 2022]

[16] Heien MH et al. Utvikling i nøkkeltal for nettselskaper. 95/2018. 2018. Available from: http://publikasjoner.nve.no/rapport/2018/rapport2018_95.pdf. [Accessed: July 13, 2022]

[17] i-Scoop. Smart Grids: Electricity Networks and the Grid in Evolution. Available from: https://www.i-scoop.eu/industry-4-0/smart-grids-electrical-grid/. [Accessed: July 13, 2022]

[18] Stadler M et al. Value streams in microgrids: A literature review. Applied Energy. 2016;**16**:980-989

[19] Beck D. Smart Energy - The Future of Power Storage. Sebastopol, California, United States: O'Reilly; 2016

[20] Ministry of Petroleum and Energy (OED). Forskrift om økonomisk og teknisk rapportering, inntektsramme for nettvirksomheten og tariffer (FOR-1999-03-11-302). (Regulation on economic and technical reporting, income limit for the limitations for the electricity network business, and tariffs)

[21] European Commission. Directive (eu) 2019/944 of the European Parliament and of the council of 5 June 2019 on common rules for the internal market for electricity and amending directive 2012/27/eu (text with eea relevance.): Pe/10/2019/rev/1," 14.6.2019. 2019. Available from: http://data.europa.eu/eli/dir/2019/944/oj. [Accessed: July 13, 2022]

[22] Kerscher S, Arboleya P. The key role of aggregators in the energy transition under the latest European regulatory framework. International Journal of Electrical Power & Energy Systems. 2022;**134**:1-16

[23] Statistics Norway. Registered vehicles. 2022. Available from: https://www.ssb.no/en/transport-og-reiseliv/landtransport/statistikk/bilparken. [Accessed: July 13, 2022]

[24] Berntzen L, Meng Q, Johannessen MR, Vesin B, Brekke T, Laur I. The aggregator as a storage provider. In: Proceedings of the 11th International Conference on Power and Energy Systems (ICPES). 2021. pp. 190-195

[25] Revankar ST. Chemical energy storage. In: Bindra H, Revanskar ST, editors. Storage and Hybridization of Nuclear Energy: Techno-Economic Integration of Renewable and Nuclear Energy. London, United Kingdom: Academic Press; 2019. pp. 177-227

[26] Smart Cities Marketplace. Electric Vehicles for Smart Cities and Communities (EV4SCC). Available from: https://smart-cities-marketplace.ec.europa.eu/action-clusters-and-initiatives/action-clusters/sustainable-urban-mobility/electric-vehicles-smart. [Accessed: July 13, 2022]

[27] Elbilforeningen. Norwegian EV Policy. Available from: https://elbil.no/english/norwegian-ev-policy/. [Accessed: July 13, 2022]

[28] BloombergNEF. Battery Pack Prices Fall to an Average of $132/kWh, But Rising Commodity Prices Start to Bite. Available from: https://about.bnef.com/blog/battery-pack-prices-fall-to-an-average-of-132-kwh-but-rising-commodity-prices-start-to-bite/. [Accessed: July 13, 2022]

Chapter 11

Heat Pump to Increase the Efficiency of a Geothermal Heating System in the City of Călimănești - Case Study

Laurentiu Constantin Lipan and Sorin Dimitriu

Abstract

Romania is one of the European countries that has a rich geothermal potential, the main uses of this resource being the spaces heating and thermal baths. Recently, concerns about increasing energy efficiency and limiting greenhouse gas emissions, have determined an increased attention to heating systems using geothermal water. These will play an important role in the future in developing sustainable energies and reducing the use of fossil fuels. The authors present a case study on a heating system in the city of Călimănești, initially using liquid fuel, which was modernized by using geothermal hot water for the preparation of the thermal agent. As geothermal water, with an initial temperature of 95°C, cannot be cooled below 50°C, the authors considered the possibility of using a heat pump to fully exploit the thermal potential of geothermal water, increase the capacity of the system and increase its energy efficiency. The implementation of the heat pump in the district heating system and the results expected to be obtained are discussed. It was considered that the heat pump works in parallel with the heating system, obtaining an increase in its capacity by approx. 60% and an increase in energy efficiency by approx. 30%.

Keywords: district heating, geothermal heat pump, geothermal heating system, energy efficiency, heat recovery, use of geothermal water

1. Introduction

Geothermal energy can be defined as the energy inside the earth, which generates geological phenomena on a planetary scale. The term used comes from the Greek words "geo" (earth) and "thermos" (heat) and is used nowadays to define that part of the earth's energy that can be recovered and used by man to generate heat and power. Geothermal energy is characterized as a renewable and sustainable energy. The character of renewable is given by its continuous production inside the earth. During the operation of natural geothermal systems, the regeneration of geothermal energy takes place by heating the geothermal water to the same time scale at which it is extracted for use. In the usual case of dry hot rocks and hot water aquifers in sedimentary basins, energy recharging is done through a slow process of thermal conduction. The sustainability of a resource depends on its initial quantity, regeneration rate and consumption rate. Obviously, consumption can be sustained if the resource is regenerated faster than it is depleted. In this context, sustainable development involves the use of the resource so that through continuous regeneration it will allow its use by future generations. The degree of sustainability of geothermal energy is still high because, compared to the volume of resources and fast regeneration rate, humanity out of convenience, uses only a small part.

Natural geothermal springs have been used for heating and thermal baths since ancient times. Archeological discoveries suggest that the earliest uses of geothermal energy took place over 10,000 years ago in North America, where autochthone populations have used the hot springs in this area for both practical and spiritual purposes. The finding that hot mineral baths ameliorate or even cure some diseases, led to the consideration of these springs as sacred, endowed by gods with magical healing powers [1].

Evidence of this has also been found in the peoples of ancient Greece and Roman Empire. At the same time, evidence was also discovered that attested to more commonplace uses: heating of living spaces, hot baths and activities related to food preparation. Evidence that geothermal energy was used for heating dates to the first century (AD), being found in the Roman city of Pompeii. However, concerns about such uses of geothermal energy were initially limited to locations where hot geothermal water was naturally accessible in the form of springs [2].

Nowadays, the direct use of geothermal energy is reported and documented in at least 88 countries. The estimated capacity to be currently installed in these countries is approx. 108,000 MW, and the annual energy consumption of approx. 1,021,000 TJ (284,000 GWh/year), being oriented towards: geothermal heat pumps (58.8%), swimming pools and thermal baths (18%), direct space heating (16%), greenhouse heating (3.5%), uses in industrial processes (1.6%), aquaculture and fish farming (1.3%), drying of cereals (0.35%), other uses (0.45%). The highest consumptions were reported in order in China, USA, Sweden, Turkey, and Japan [3].

According to data reported at the World Geothermal Congress in 2020 (WGC 2020), only in recent years, 2015-2020, there has been an increase in the amounts of energy used from geothermal sources by approx. 27%. Concerns about the production of electricity in ORC installations that take heat from geothermal waters have also intensified, with five countries installing production capacity for the first time: Belgium (0.8 MW), Chile (48 MW), Croatia (16.5 MW), Honduras (35 MW) and Hungary (3 MW). Approximately 2650 wells were drilled in 42 countries and approx. $ 22.3 billion has been invested in geothermal projects [4].

Geothermal energy resources are located over a wide range of depths and can be in the form of hot water, steam, or hot rocks. Hot water aquifers with temperatures between 60°C and 100°C are the most suitable applications for space heating and agricultural systems. For these aquifers to be commercially interesting, they must be located at depths up to 2000 - 3000 m and have a temperature of at least 60°C. Areas with hot water under pressure with higher temperatures and areas with hot rocks are suitable as a source for electricity generation [5].

Traditionally, the heating of homes, office buildings and commercial spaces has been done with the help of local heat sources: stoves, fireplaces and hot water boilers using different types of fuels. These systems not only have low energy efficiency but are also powerful generators of carbon dioxide as well as various polluting suspensions. For urban areas with high population density, all studies conducted at national and international level have led to the conclusion that from the point of view of energy efficiency and environmental protection, district heating (DH) systems are advantageous [6].

The multiple and obvious advantages of district heating are: high energy efficiency; the possibility of using several types of fuels; use of residual energy resulting from industrial processes (hot water, steam); use of renewable resources (solar energy, geothermal water, biomass, biofuels, household waste and other combustible waste); simple operation by the consumer, which is not involved in fuel supply activities, maintenance and operation supervision; consumer safety, compared to individual sources; reduced pollution, by placing thermal energy sources outside the living area and achieving a low level of pollutant emissions and greenhouse gases; the possibility of applying local investment policies in the field of energy efficiency and improving the quality of the environment.

Despite all these major advantages, compared to the alternative of individual heating, the consumer connected to a DH system also faces a certain degree of limitation of thermal comfort, determined by how the system can respond to variable loads or to operate economically under load limitation. Nevertheless, the DH system solution provides the necessary heating and hot water at prices less than or equal to those offered by individual alternative solutions. District heating is a suitable solution for all sizes of networks, from a few buildings to the neighborhood or city level. This has strengthened of DH system position in many European countries in recent years [7, 8].

In the current conditions in which the amplification of global warming imposes firm and hard measures for limiting the greenhouse gas emissions by restricting the use of traditional fuels, the integration of geothermal resources in DH systems is a major requirement. Geothermal heating systems not only reduce or eliminate carbon dioxide emissions but, by eliminating the consumption of conventional fuels, provide consumers with heat at much lower prices, especially in the current context of the gas and oil crisis. Geothermal heat resources have been used in district heating systems for many years in Iceland and France. Other such installations appear in Germany, Hungary, Italy, Romania, Belgium, and the United Kingdom. During 2014, 30 PJ of heat were supplied from geothermal sources worldwide, of which 7.3 PJ in the European Union. However, these reserves appear to be somewhat underestimated, according to reports from the European GeoDH project. Given that about a quarter of the European Union's population lives in urban areas where geothermal energy could be extracted and used, future geothermal heating systems appear as an efficient solution to the current problems of thermal energy supply [9].

2. Geothermal heating system in the city of Călimănești, Romania

Romania has an important potential of geothermal energy sources, on its territory being identified several areas in which the geothermal potential is estimated to allow economic applications. They are located on an extensive area in western plain of Romania, in the middle of country, on the Olt River Valley north of Râmnicu Vâlcea, in the northern part of Bucharest city and south of Brăila city. **Figure 1** shows the locations of these geothermal reservoirs in Romania. Geological surveys conducted before 1990 showed that the known geothermal potential in Romania is about 10PJ/year, of which only about two thirds is exploited. The approximately 80 functional wells can produce an annual amount of thermal energy of about 7PJ [10].

Figure 1.
The distribution of geothermal reservoirs on the Romanian territory [10].

In Romania, the temperature of "low enthalpy" geothermal sources, with exploitation by drilling-extraction is between 25°C and 60°C, and 60–125°C for the "mesothermal". The economic drilling and extraction limit for geothermal waters was agreed for the depth of 3300 m and was reached in some areas of Romania, such as the Bucharest North - Otopeni geothermal basin and certain perimeters in Snagov and Balotești localities. Romania was ranked as the third country in Europe, after Greece and Italy, for its very high geothermal potential [11].

More than 80% of the wells are exploited artesian, 18 wells require anti-scaling chemical treatment, and 6 are used for reinjection. The main direct uses of the geothermal energy are space and district heating; bathing; greenhouse heating; industrial process heat; fish farming and animal husbandry [11].

This chapter presents a case study. The authors present a solution to increase the energy efficiency of the district heating system of the city of Călimănești, which works having as main source of heat the geothermal water extracted from the aquifer of Călimănesti – Căciulata - Cozia perimeter, located on the Olt River Valley. In this area, the geothermal water is provided by three drillings having more than 3000 m in depths, located on the right side of the Olt River, at about 1-2 km one from each other as presented in the **Figure 2**. The three existing drillings stand out deposits of medium enthalpy geothermal water (the temperature at the exit of the well is 92 ... 95°C). The available flow volume of the three wells is 50.4 l/s, equivalent to a thermal potential of 13.2 MW, when the geothermal water is cooled to 30°C [12].

The drilling located in vicinity of Căciulata and Cozia localities (borehole 1008 and 1006) are used only for local heating needs. The geothermal water feeds a group of hotels and SPA treatment units, for heating, domestic hot water supply and thermal pools. The high thermal potential of the geothermal water leads to its direct

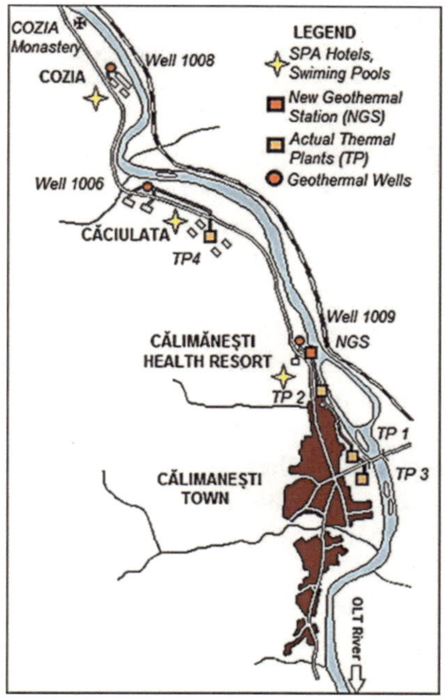

Figure 2.
Olt Valley working geothermal perimeter [12].

exploitation. In the cold season, the geothermal water (having a temperature of 92 ... 95°C) is cooled in a plate heat exchanger, producing the thermal agent (TA) for the district heating system. A second heat exchanger produces domestic hot water (DHW). The geothermal water, cooled in the two heat exchangers, feeds the thermal pool, after that being discharged in the Olt River at a temperature of about 30°C. In the warm season, the mass flow extracted is reduced, only the heat exchanger for domestic hot water and thermal pool being in use.

The third drilling is situated at 1,2 km from Călimăneşti, providing a volume flow of 18 l/s at the same temperature values 92 ... 95°C. This locality, beside the tourists which are staying in hotels, has about 8600 permanent habitants; 15% of the habitants are living in apartments connected to a centralized system for thermal energy supply. In the cold season of 2019-2020, 461 apartments and residential houses, 9 public institutions and 47 economic operators were branched to DH system, which must ensure a thermal need of about 3500 kW for heating and about 500 kW for DHW supply (for the conventional climatic parameters) [13]. The DH system was initially designed with three thermal units, equipped with hot water boilers using light liquid fuel. The geothermal water from the nearby well was initially used only for the thermal energy supply of the SPA treatment units and thermal pools. The project of the DH system feeding with geothermal energy was started in 2002 with internal financing and was later supported by European funds. Initially, the project included all the three wells (1006, 1008 and 1009) to provide the centralized heating of Călimănesti town. Later, it was utilized only the available water from the well 1009, situated in vicinity of town. The available volume flow is of 18 l/s, from which about 8 l/s is utilized by a SPA center and a hotel; the rest of volume flow (about 10 l/s) being used in the DH system of Călimănesti. To use the geothermal water into the DH system, a geothermal heating station was built just near the geothermal well. The geothermal water produces, by using plate heat exchangers, the primary thermal fluid for the DH system, having a temperature of about 85°C. This primary thermal fluid serves to partially cover the heating demand and to completely cover the DHW preparation.

Figure 3 shows the drilling 1009 and the geothermal station that prepares the heating agent for the DH system. The geothermal water from the Olt Valley aquifer has a high combustible gas content, containing over 90% methane. As a result, before being introduced into the heat exchangers of the geothermal station, it is degassed, the collected gases being discharged into the atmosphere. **Figure 4** shows the actual operating diagram of the geothermal station.

The primary thermal agent from the DH system of Călimănești town is returned to the geothermal station with a temperature around 40°C. For this reason, the geothermal water extracted from the drilling well, with a temperature of 95°C, cannot be cooled, in the heat exchangers of geothermal station, to a temperature below of about 50°C. Because the exploitation of geothermal aquifer is artesian, this water is discharged directly into the Olt River after a cooling until 30°C, from reason of aquatic environment protection. The cooling and discharge into the Olt River of the waste geothermal water, represents a heat loss of approx. 1/3 of its full thermal potential. The authors examined the possibility of recovering this heat loss by implementing in the thermal agent preparation circuit a heat pump that uses as heat source the waste geothermal water discharged from the geothermal station, cooling it from 50–30°C. In this way, the entire thermal potential of geothermal water is used, while also increasing the capacity of the DH system by growing the flow of thermal agent produced. Waste geothermal water can be discharged directly into the Olt River, without any negative impact on the aquatic environment. The functional diagram of the geothermal station, coupled with a mechanical vapor compression heat pump, is presented in **Figure 5**.

Figure 3.
The borehole 1009 and geothermal station (source: photo of the authors).

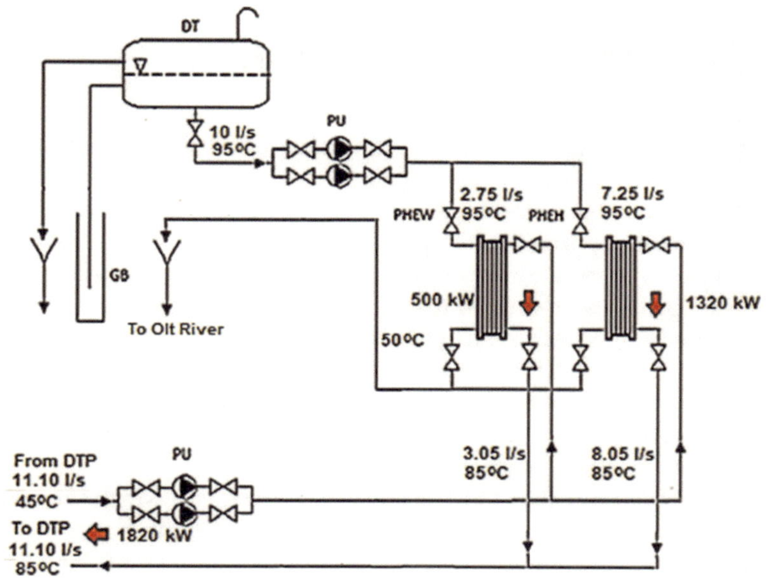

Figure 4.
The actual operating diagram of the geothermal station [12]. GB – Geothermal borehole; DT – Degassing tank; PU – Pumping units; DTP – Distribution thermal points; PHEH – Plate heat exchanger for heating; PHEW – Plate heat exchanger for DHW.

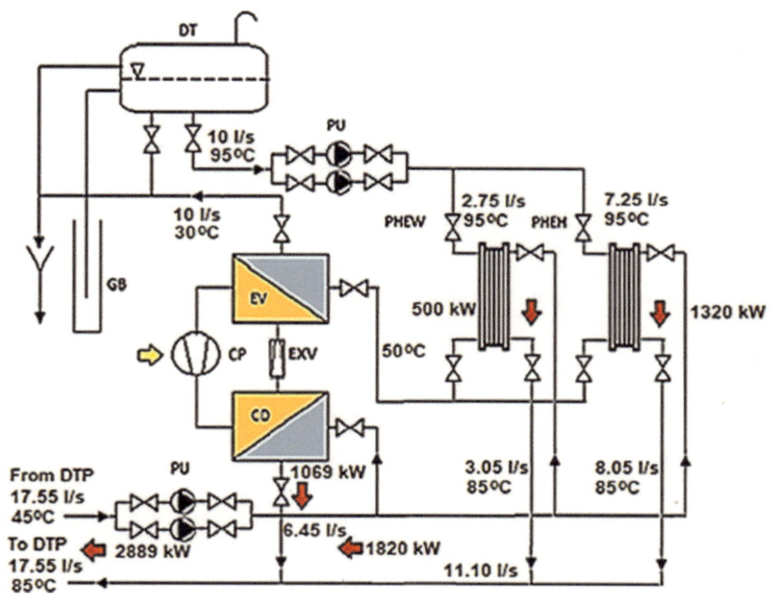

Figure 5.
The functional diagram of the geothermal station with heat pump (source: the drawing of authors). CP- compressor; EXV – Expansion valve; CD – Condenser; EV – Evaporator.

3. The geothermal heat pump

The idea of introducing a heat pump in the functional diagram of the geothermal station was to connect it in parallel with the heat exchangers that prepare the heat sent to the network of the DH system, supplementing its flow. The flow of geothermal water available from the wellbore is 10 l/s, which allows the coverage of a thermal load of 1820 kW, the geothermal water being cooled only to 50°C. Using the heat pump for further cooling of the geothermal water until a temperature of 30°C, an additional thermal power of 1069 kW can be obtained, estimating a value of the heat pump COP of 4,5. This result shows that the capacity of the district heating system is increased by about 60%. For geothermal wastewater to be used as a heat source, if it is cooled to 30°C, the refrigerant vaporization temperature must be around 25°C. As the temperature of the agent supplied to the heating system by the geothermal station is 85°C, the refrigerant used by the heat pump must have a condensation temperature of at least 90°C. Consequently, it is necessary to use high temperature working agents with a critical temperature above 100°C. In addition to this requirement, the refrigerant used must have an environmental impact in accordance with the provisions of the EU Regulation on fluorinated greenhouse gases and have thermodynamic properties to achieve the highest possible coefficient of performance.

Most of the agents shown in **Table 1** are "wet agents", for which the end of dry saturated vapors compression, from the evaporator, falls within the saturation domain. To avoid this phenomenon, a superheat of the vapors must be introduced at the compressor suction, of at least 10 degrees. The cooling of the heat pump condenser being carried out with the returned agent from the DH system, having a temperature of 45°C, a subcooling of refrigerant until a temperature around 60°C, is possible. **Figure 6** shows how to perform the processes of the condensate subcooling and superheating of cold vapors at the compressor suction. The cooling is carried out either in a separate heat exchanger or in the final part of the condenser and the superheating of the vapors, by introducing an internal regenerative heat exchanger in the operation scheme of the heat pump.

Subcooling of the condensate has the effect of increasing the specific thermal load of the condenser, respectively of the energy efficiency (*COP*). With the agents presented in **Table 1**, the authors performed the analysis of the thermodynamic cycle of the heat pump, to choose the most suitable agent for the imposed operating conditions:

Refrigerant	Group	p_{sat} at 20°C (bar)	t_{crit} (°C)	ODP	GWP/CO_2	Safety class
R134a	HFC	6.654	101.0	0	1430	A_1
R245fa	HFC	1.486	154.0	0	1030	B_1
R236fa	HFC	2.719	124.9	0	9810	A_1
R152a	HFC	5.979	113.2	0	138	A_2
R600a	HC	3.507	134.9	0	3	A_3
R600	HC	2.433	154.2	0	4	A_3
RE170	HC	6.908	127.2	0	3	A_3
R515b*	HFO	4.974	108.9	0	293	A_1
R1234ze(E)	HFO	4.985	109.4	0	1	A_2L
R1233zd(E)	HFO	1.298	165.5	0	1	A_1

Note: HFC = hydrofluorocarbons; HC = Hydrocarbons; HFO = Hydrofluoro-olefins.
**Azeotropic blend R1234ze/R227ea (91,1%/8,9%).*

Table 1.
The properties of a few agents used for high temperature heat pumps [14].

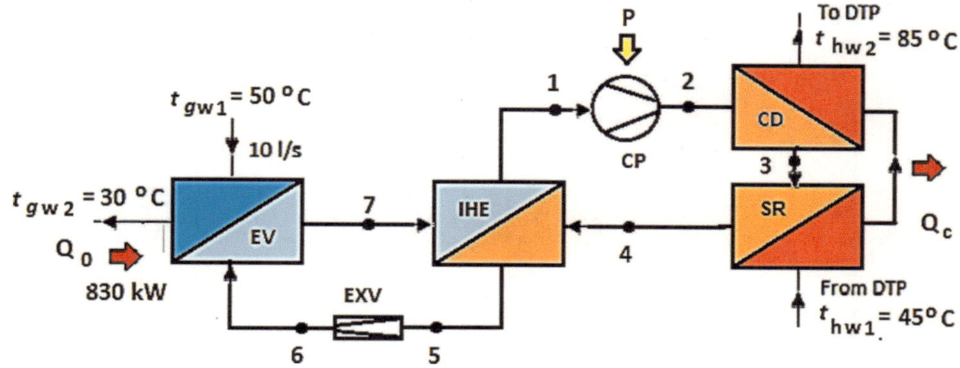

Figure 6.
The operating diagram of the heat pump installation (source: the drawing of authors). EV – Evaporator; IHE – Internal heat exchanger; CP – Compressor; CD – Condenser; SR – Subcooler; EXV – Expansion valve; DTP – Distribution thermal point.

- heating of the thermal agent in the condenser: from 45–85°C;
- condensate subcooling temperature (t_{sc}): 60°C;
- cooling of the geothermal water in the evaporator: from 50–30°C;
- vapor overheating at the compressor suction (Δt_{sh}): 10 degrees;
- pinch point heat exchangers (Δt_{gap}): around 5 degrees;
- the isentropic efficiency of the compressor (η_c): 0.8.

For all the considered agents, according to these conditions, the condensation temperature was chosen $t_c = 90°C.$, and the vaporization temperature $t_v = 25°C$

The thermodynamic cycle, in the p-h diagram, is presented in **Figure 7**.

Saturation pressures corresponding to these temperatures depend on the agent considered:

$$p_v = p_{sat}(t_v); \quad p_c = p_{sat}(t_c) \tag{1}$$

The superheating of the vapors sucked by the compressor is carried out with the help of the hot condensate having a temperature of 60°C, by means of the regenerative heat exchanger. Since in this heat exchanger, the flow of the hot fluid is equal to that of the cold fluid, the thermal balance of the appliance has the form:

$$h_4 - h_5 = h_1 - h_7 \tag{2}$$

specifying the enthalpy of condensate at the inlet to the expansion valve.

The state parameters in the characteristic points of the operating scheme in **Figure 6**, respectively of the thermodynamic cycle in the p-h diagram in **Figure 7**, were determined using the EES software. The **Table 2** shows the calculation algorithm of the state parameters of the cycle.

According to the thermodynamic cycle in **Figure 7**, the specific energy parameters of the installation are:

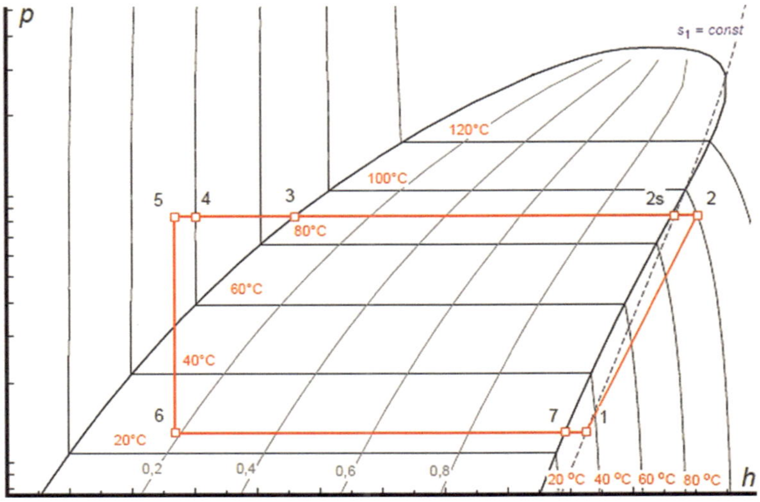

Figure 7.
The thermodynamic cycle of the heat pump installation, in p-h diagram (source: generated in EES software by authors).

State No.	State description	Known state parameters	Functions for determining state parameters
1	Superheated vapors at the compressor suction	$p_1 = p_v$ $t_1 = t_v + \Delta t_{sh}$	$h_1 = enthalpy(p_v, t_1)$ $s_1 = entropy(p_v, t_1)$
2 *s*	Superheated vapors at the end of isentropic compression	$p_{2s} = p_c$ $s_{2s} = s_1$	$t_{2s} = temperature(p_c, s_{2s})$ $h_{2s} = enthalpy(p_c, t_{2s})$
2	Superheated vapors at the end of real, irreversible compression	$p_2 = p_c$ $h_2 = h_1 + (h_{2s} - h_1)/\eta_c$	$t_2 = temperature(p_c, h_2)$ $s_2 = entropy(p_c, h_2)$
3	Liquid at saturation	$p_3 = p_c$ $t_3 = t_c$ $x_3 = 0$	$h_3 = enthalpy(p_c, x_3)$ $s_3 = entropy(p_c, x_3)$
4	Subcooled liquid	$p_4 = p_c$ $t_4 = t_{sc}$	$h_4 = enthalpy(p_c, t_{sc})$ $s_4 = entropy(p_c, t_{sc})$
5	Subcooled liquid at the inlet of the expansion valve	$p_5 = p_c$ $h_5 = h_4 + (h_1 - h_7)$	$t_5 = temperature(p_c, h_5)$ $s_5 = entropy(p_c, h_5)$
6	Two-phase mixture after expansion valve	$p_6 = p_v$ $t_6 = t_v$ $h_6 = h_5$	$s_6 = entropy(p_v, h_6)$ $x_6 = quality(p_v, h_6)$
7	Dry saturated vapors at the exit of the evaporator	$p_7 = p_v$ $t_7 = t_v$ $x_7 = 1$	$h_7 = enthalpy(p_v, x_7)$ $s_7 = entropy(p_v, x_7)$

Table 2.
Algorithm for determining the state parameters in EES software.

specific thermal load of the condenser:

$$|q_c| = h_2 - h_4 \text{ [kJ/kg]} \quad (3)$$

specific cooling power of the evaporator:

$$q_v = h_7 - h_6 \ [\text{kJ/kg}] \tag{4}$$

specific mechanical compression work:

$$|l_{cp}| = h_2 - h_1 \ [\text{kJ/kg}] \tag{5}$$

The heat flow taken up by the heat pump evaporator depends on the available geothermal water flow $\dot{V}_{gw}$ [m^3/s] and the temperature up to which the water can be cooled:

$$\dot{Q}_v = \dot{V}_{gw}\rho_{gw}\left(h_{gw1} - h_{gw2}\right) \ [\text{kW}] \tag{6}$$

where ρ_{gw} [kg/m^3], the density of waste geothermal water, was considered at its average temperature $t_{gw} = 0.5\left(t_{gw1} + t_{gw2}\right)$. The flow of refrigerant is determined by the heat flow recovered from waste geothermal water:

$$\dot{m} = \frac{\dot{Q}_v}{q_v} \ [kg/s] \tag{7}$$

and establishes the heat flow given to the condenser and respectively the theoretical power required to drive the compressor:

$$|\dot{Q}_c| = \dot{m}\ |q_c| \ [\text{kW}] \tag{8}$$

$$P_{cp} = \dot{m}\ |l_{cp}| \ [\text{kW}] \tag{9}$$

resulting in COP (energy efficiency)

$$COP = \frac{|\dot{Q}_c|}{P_{cp}} \tag{10}$$

The efficiency of the Carnot cycle carried out between the two heat sources of the thermodynamic cycle, the hot source, and the cold source, has the expression:

$$COP_C = \frac{T_{HS}}{T_{HS} - T_{CS}} \tag{11}$$

wherein the temperatures of the two heat sources, T_{HS} and T_{CS}, are the mean thermodynamic temperatures of the hot water produced at the condenser and respectively of the geothermal water from which the heat is extracted at the evaporator.

The mechanical work of the reversible Carnot cycle has the significance of the minimum mechanical work necessary to achieve heat transfer from cold to hot source. In the case of a real irreversible cycle, carried out between the same heat sources, the internal and external irreversibility, determine a destruction of the available energy, determining the increase of the mechanical work consumption. The most crucial parameter for the thermodynamic evaluation of a cycle, from this point of view, is the exergetic efficiency (carnotic efficiency):

$$\eta_{ex} = \frac{|l_C|}{|l_{cycle}|} = 1 - \left(\sum \overline{\pi}_{ir,int} + \sum \overline{\pi}_{ir,ext}\right) = \frac{COP}{COP_C} \tag{12}$$

wherein $\sum \overline{\pi}_{ir,int}$ is the sum of the rates of the losses caused by the internal irreversibility of the cycle and $\sum \overline{\pi}_{ir,ext}$, the sum of the rates of the losses caused by its external irreversibility.

To determine which of the working agents shown in **Table 1** is more suitable for this heat pump installation, in terms of properties and performance, were determined the state parameters for each point of the cycle in **Figure 7**, according to the calculation algorithm presented in **Table 2**. The calculus was carried out in the same initial conditions for each refrigerant. The drive power of the compressor and the heat flow yielded to the condenser were also calculated, resulting in the performance coefficient of the installation. The reference Carnot cycle was considered between the average thermodynamic temperatures of the geothermal water in the evaporator and respectively of the thermal agent in the condenser.

The results obtained are presented in **Table 3** and in **Figure 8**.

All refrigerants that have been analyzed are in the category of those that do not affect the ozone layer, with zero ODP potential. In terms of performance, R245fa, R600 and R1233ze(E) refrigerants for which energy efficiency (COP) has the highest values are noted (**Figure 8**).

R245fa is a colorless, one-component fluid of class HFC (1,1,1,3,3-Pentafluoropropane, $C_3H_3F_5$) which may replace the use of HCFC R123 and R11. R245fa is among the HFC refrigerants that do not deplete the ozone layer (ODP = 0) but have a significant global warming potential (GWP = 1030). R245fa is used primarily as a blowing and insulation agent for plastic foam insulation, but also as an industrial air conditioning refrigerant, heat recovery systems and high energy recovery systems. HFC-245fa is listed as non-toxic and non-flammable, but exposure to high levels of R245fa can lead to heart sensitization and eye irritation. It falls into safety class B_1 [15].

Refrigerant	q_v (kJ/kg)	$\lvert q_c \rvert$ (kJ/kg)	$\lvert l_{cp} \rvert$ (kJ/kg)	$\dot{m}$ (kg/s)	$\lvert \dot{Q}_c \rvert$ (kW)	P_{cp} (kW)	*COP*	η_{ex}
R134a	136.0	178.3	42.29	6.107	1089	258.3	4.216	0.305
R245fa	152.0	195.9	43.92	5.464	1070	240.0	4.460	0.322
R236fa	109.3	141.8	32.50	7.559	1078	247.0	4.363	0.315
R152a	225.4	294.1	68.71	3.685	1084	253.2	4.280	0.309
R600a	257.1	333.3	76.12	3.231	1077	245.9	4.379	0.316
R600	289.8	373.9	84.14	2.866	1072	241.2	4.444	0.321
RE170	328.5	426.9	98.40	2.528	1079	248.8	4.338	0.313
R515b	121.6	159.0	37.40	6.831	1086	255.5	4.251	0.307
R1234ze(E)	125.9	164.2	38.29	6.597	1083	252.6	4.288	0.310
R1233zd(E)	155.7	200.4	44.62	5.335	1069	238.0	4.491	0.325

Note: Operating parameters, same for all refrigerants:
Condensation temperature: 90 °C.
Vaporization temperature: 25°C.
Superheating at the compressor suction (Δt_{sh}): 10°.
Thermodynamic mean temperature of the geothermal water: 40.55°C.
Thermodynamic mean temperature of the thermal agent: 64.55°C.
Heat flow taken from the evaporator: 830.60 kW.
Efficiency of the reference Carnot cycle: 13.84.

Table 3.
The performances of the heat pump, for different refrigerants.

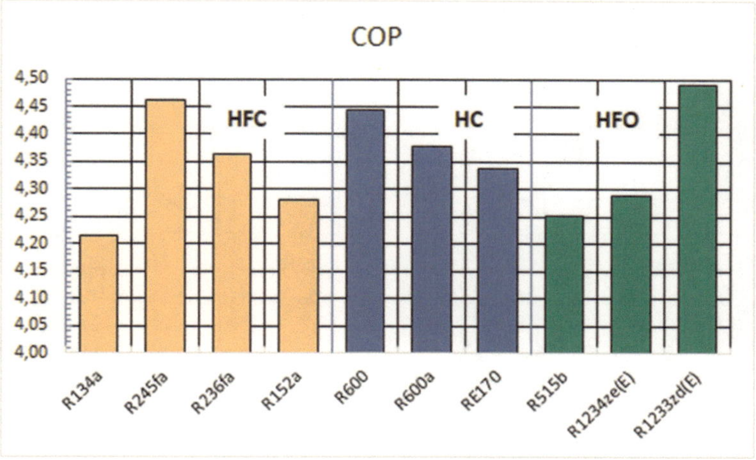

Figure 8.
The COP of the heat pump installation, for different agents (source: results from ***Table 3****).*

R600 refrigerant is a natural refrigerant of class HC (butane, C_4H_{10}) suitable for many refrigeration applications. R600 refrigerant is environmentally friendly and efficient. R600 is a commonly used refrigerant with low environmental impact and good thermodynamic performance. The R600 has low power consumption in refrigeration applications and is compatible with many different lubricants. The R600 is suitable for a variety of tasks in industrial, commercial, and household refrigeration, such as kitchen refrigerators and professional freezers and desks, as well as refrigeration and freezing appliances. R600 is classified as a highly flammable class A_3 refrigerant and is subject to strict regulations regarding the amount of agent charged into installations [16].

R1233zd(E) refrigerant is a recent agent, chemically a hydrochlorofluoroolefin (HFO, CF_3-CH=CClH). Although this refrigerant is an HCFC and therefore carries chlorine that affects the ozone layer, it is not on the list of fluorinated greenhouse gases that will be eliminated, because the life in the atmosphere is very short (26 days). The properties of this agent are close to the characteristics of the ideal refrigerant: adequate operating pressures, zero GWP potential, zero ODP potential, non-flammable, non-toxic, and adequate volumetric capacity. R1233ze(E) is classified as safety class A_1 [15].

Based on the energy performance and physical properties of the selected agents, it was decided that the most suitable agent, to be used in the studied heat pump, is R1233ze(E) refrigerant.

For the refrigerant considered the most suitable, the way in which the energy efficiency (COP) and the heat flow produced at the condenser depend on the subcooling temperature and on the superheating of the vapors at the compressor suction, respectively, was studied. The results are shown in the **Figure 9**.

It is found that the greatest influence upon the heat pump performances has the subcooling degree of the condensate, while the influence of the degree of overheating at the compressor suction is practically insignificant. For this reason, the minimum value of 10 degrees was considered for the degree of overheating, which ensures that the state corresponding to the compressor discharge does not enter in the wet area and was investigated the way in which only the degree of subcooling influences the performance of the heat pump. The results are shown in **Figure 10**. It is found that the

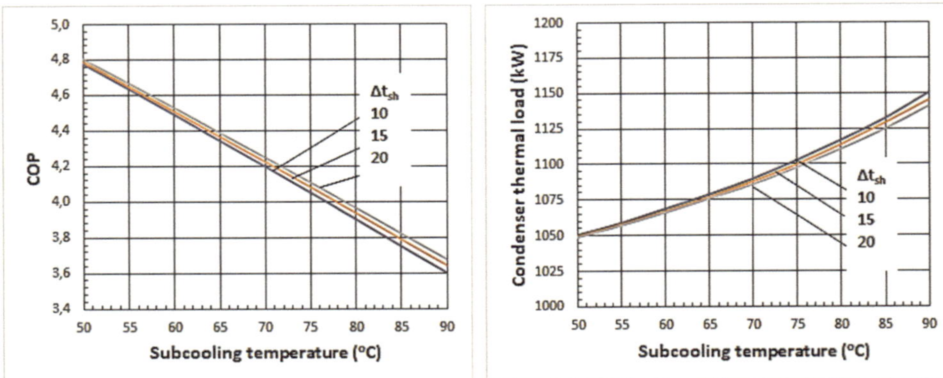

Figure 9.
Modification of the coefficient of performance and of the thermal flux produced at the condenser, depending on the subcooling temperature and the degree of overheating (source: simulation in EES software by authors).

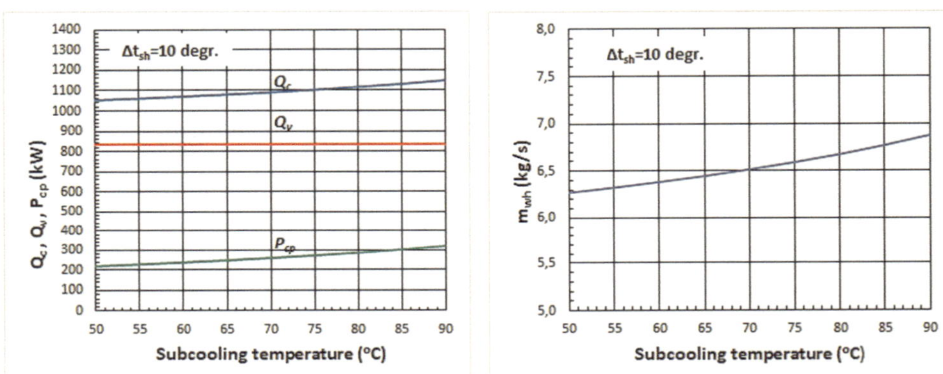

Figure 10.
Modification of the thermal flux produced at the condenser, the cooling power, the compressor driving power and the flow of the hot agent, depending on the subcooling temperature, for vapors superheating of 10 degrees (source: simulation in EES software by authors).

lower the cooling temperature, the higher the specific cooling power and the lower the required refrigerant flow. As a result, the power required to drive the compressor is reduced. Although the specific thermal load of the condenser increases, the decrease of the refrigerant flow has a stronger influence and as a result the heat flow yielded at the condenser decreases, causing a slight decrease of the heat flow provided in the district heating network. The decrease in the drive power of the compressor has a greater influence on the energy balance, so that overall, the coefficient of performance increases.

4. The coverage of the HD system

The city of Călimănești is part of the climate zone III of Romania, for which the conventional outdoor temperature for the cold season is $t_{ec} = -18°C$ [16]. For residential and tertiary buildings, the conventional interior design temperature may be considered $t_{ic} = +20°C$, as [17]. In these conditions, considering the usual thermal characteristics of the buildings and the need of DHW, the maximum thermal load of the DH system of the Călimănești city was evaluated as follow:

maximum thermal load for heating: $\dot{Q}_{heat}^{max} = 3500$ kW;maximum thermal load for DHW preparation: $\dot{Q}_{dhw}^{max} = 500$ kW.

Assuming that the heat transfer characteristics are unchanged in relation with outdoor temperature, the thermal load of the DH system, which must be performed by the geothermal station, varies linearly with the temperature of the external environment t_e [°C], according to the relation:

$$\dot{Q}_{GTS} = \dot{Q}_{heat}^{max} \frac{t_{ic} - t_e}{t_{ic} - t_{ec}} + \dot{Q}_{dhw}^{max} [kW] for t_e \leq 10°C \tag{13}$$

The variation of the thermal load in relation to the outdoor temperature is shown in **Figure 11**. It is considered that the district heating system is put into operation when the mean daily temperature of the external environment is lower than 10°C, producing thermal agent both for the preparation of domestic hot water and for heating. For mean daily outdoor temperatures above 10°C, the system prepares agent for domestic hot water only.

In the cold season 2020-2021, the variation of the mean daily outdoor temperature and necessary thermal load for heating system in the Călimănești zone, is shown in the **Figure 12** [18]. In recent years, due to global warming, mean temperatures during the cold season have been higher than usual. It can be noted that mean outdoor temperatures below 0°C was recorded in the area for a few days, only in the second half of January and in February, the thermal load during this period being about 70% of the maximum load calculated, based on the conventional temperatures. The system was started on Octomber 01, 2020 and was stopped on April 30, 2021.

The operation diagram of the geothermal station coupled with the heat pump is shown in the **Figure 13**. By cooling the geothermal water with $\Delta t_1 = 45°C$ (from 95–50°C) the heat flow introduced in the system is:

$$\dot{Q}_{he} = \dot{m}_{gw} c_w \Delta t_1 = \dot{m}_{he} c_w \Delta t \text{ [kW]} \tag{14}$$

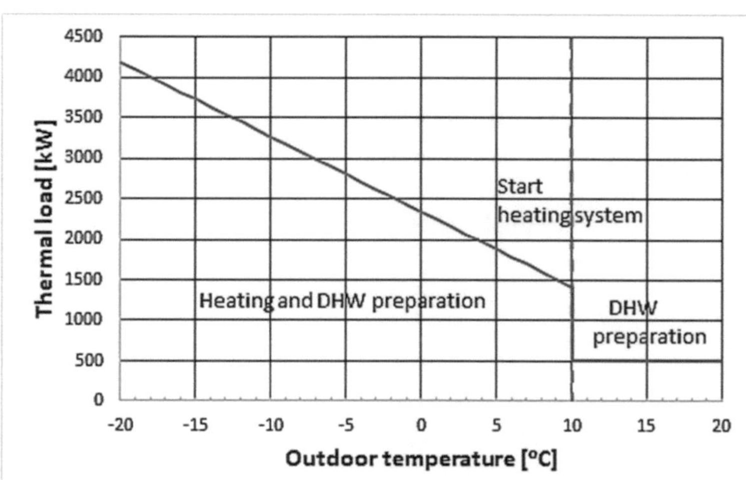

Figure 11.
The thermal load of district heating system in relation with outdoor temperature according to Eq. (13).

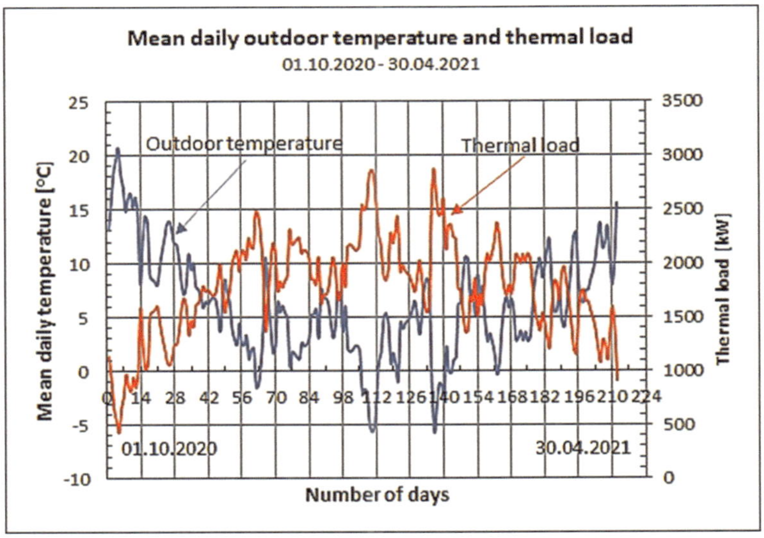

Figure 12.
Variation of mean daily outdoor temperature in area and thermal load, in the cold season of 2020-2021 [18].

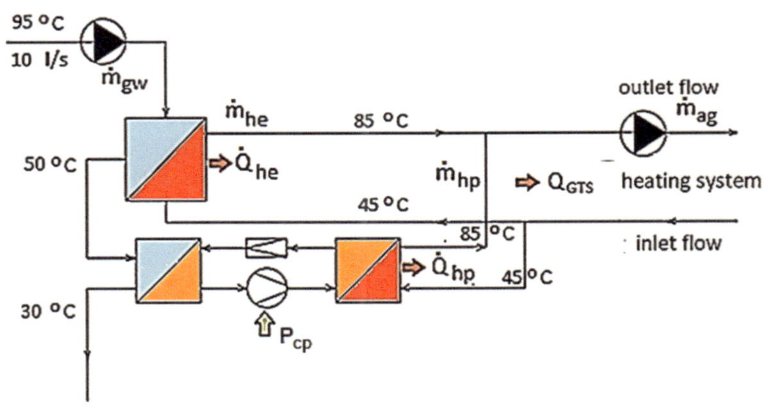

Figure 13.
Operation diagram of the geothermal station (source: the drawing of authors).

and by recovering the residual heat from the geothermal water, the heat flow introduced by the heat pump into the system is:

$$\dot{Q}_{hp} = \dot{m}_{gw} c_w \Delta t_2 \frac{COP}{COP - 1} = \dot{m}_{hp} c_w \Delta t \ [kW] \tag{15}$$

where, $\Delta t_2 = 20°C$ (from 50–30°C) is the degree of geothermal water cooling in the heat pump evaporator, COP is energy efficiency, and $\Delta t = 40°C$ (from 85–45°C) the temperature difference between the outlet and the inlet of the thermal agent sent into the district heating system.

The two expressions determine the total heat flow that the geothermal station coupled with the heat pump, introduces into the district heating system:

$$\dot{Q}_{GTS} = \dot{m}_{gw}c_w\left(\Delta t_1 + \Delta t_2\frac{COP}{COP-1}\right) = \left(\dot{m}_{he} + \dot{m}_{hp}\right)c_w\Delta t \quad [kW] \tag{16}$$

For the maximum available flow of geothermal water of 10 l/s, the maximum heat flow delivered to the DH system, according to relation (16), is $\dot{Q}_{GTS}^{max} = 2889$ kW of which 1820 kW directly from geothermal water and 1069 kW from the heat recovered by means of the heat pump. For the operating conditions of the heating system, the energy efficiency of the heat pump was considered *COP* = 4.5, according to cycle analyze. The external limit temperature up to which the system can operate only with the thermal energy produced in the geothermal water heat exchanger is:

$$t_e^{gw} = t_{ic} - \frac{\dot{Q}_{he}^{max} - \dot{Q}_{dhw}^{max}}{\dot{Q}_{heat}^{max}}(t_{ic} - t_{ec})[°C] \tag{17}$$

and the external limit temperature up to which the system can operate coupled with heat pump is:

$$t_e^{gw+hp} = t_{ic} - \frac{\dot{Q}_{he}^{max} + \dot{Q}_{hp}^{max} - \dot{Q}_{dhw}^{max}}{\dot{Q}_{heat}^{max}}(t_{ic} - t_{ec})[°C] \tag{18}$$

According to the operating conditions of the heating system, for the limit temperatures specified by relations (16) and (17) the following values resulted for the limit temperatures: $t_e^{gw} = 5.7°\mathbf{C}$ and $t_e^{gw+hp} = -5.5°\mathbf{C}$. *For outdoor temperature lower than* t_e^{gw+hp}, gas hot water boilers must also be started. The **Figure 14** shows how the geothermal station can cover the thermal load of the district heating system, depending on the level of the outside temperature.

The adjustment of the geothermal station functioning, so that the heat flow produced can cover the thermal load determined by the outdoor temperature, can be done by changing the flow of the thermal agent sent in the network, with the constant maintenance of its temperature (quantitative regulation) or with the constant maintenance of

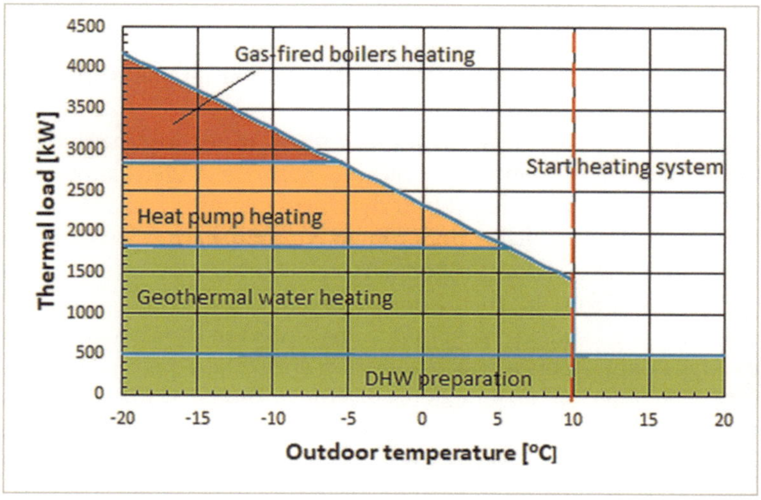

Figure 14.
The thermal load coverage in relation with outdoor temperature (source: the drawing of authors).

the flow of the agent sent in the network and the change of its temperature (qualitative regulation). Only quantitative adjustment has been considered in this discussion.

The heat flow required for the DH system needs a water flow to be taken from geothermal borehole:

if $t_e \geq 10°\mathbf{C}$ (production of DHW only)

$$\dot{m}_{gw} = \frac{\dot{Q}_{dhw}^{max}}{c_w \Delta t_1} [kg/s] \tag{19}$$

if $10°C > t_e \geq 5.7°\mathbf{C}$ (production of DHW and heating with geothermal water only)

$$\dot{m}_{gw} = \frac{\dot{Q}_{heat}^{max} \frac{t_{ic}-t_e}{t_{ic}-t_{ec}} + \dot{Q}_{dhw}^{max}}{c_w \Delta t_1} [kg/s] \tag{20}$$

if $5.7°C > t_e \geq -5.5°\mathbf{C}$ (production of DHW and heating with geothermal water and heat pump)

$$\dot{m}_{gw} = \frac{\dot{Q}_{heat}^{max} \frac{t_{ic}-t_e}{t_{ic}-t_{ec}} + \dot{Q}_{dhw}^{max}}{c_w \left(\Delta t_1 + \Delta t_2 \frac{COP}{COP-1}\right)} [kg/s] \tag{21}$$

The HD system can cover the required thermal load using only geothermal water, up the outdoor temperature of +5.7°C, when the maximum geothermal water flow of 10 l/s (9.62 kg/s) is extracted from the borehole. Below this temperature, the DH system coupled to the heat pump can covers the required thermal load up to temperature of −5.5°C, when the flow of extracted geothermal water is also maximum. For lower outdoor temperatures, the thermal load is supplemented by the commissioning of hot water boilers with gaseous fuel.

If the temperature of the thermal agent sent to the district heating system is kept constant, its flow rate depending on the temperature of the external environment is expressed:

$$\dot{m}_{ta} = \dot{m}_{he} + \dot{m}_{hp} = \frac{\dot{Q}_{heat}^{max} \frac{t_{ic}-t_e}{t_{ic}-t_{ec}} + \dot{Q}_{dhw}^{max}}{c_w \Delta t} [kg/s] \tag{22}$$

The variation of these flows in relation to the outdoor temperature is shown in the **Figure 15**.

Coupling the geothermal station with a heat pump to recover the thermal energy of the wastewater discharged from the geothermal heat exchangers, allows to increase the flow of thermal agent introduced into the district heating system and cover about 70% of its maximum thermal load. Under these conditions, the district heating system can ensure the thermal comfort of consumers only up to an outside temperature of around −6°C (**Figure 16**).

Examining the way in which the climatic conditions in the area have manifested in recent years, by coupling the geothermal station with a heat pump, the maximum available flow of geothermal water can ensure the coverage of the entire thermal load of the heating system, without the use of gas-fired hot water boilers. Due to the global warming phenomenon, the mean daily temperatures during the cold season were higher than the usual temperatures for this period.

As can be seen in the **Figure 16**, according to the weather archive in the area, during the cold period of the 2020-2021 season, in just a few days the average outdoor

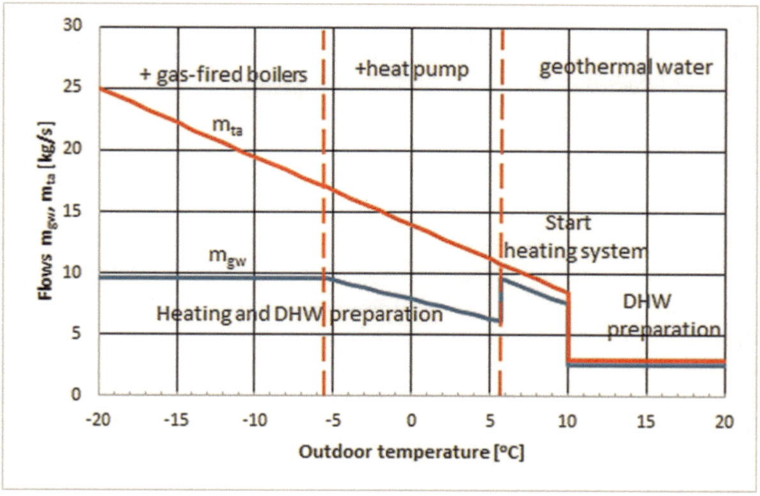

Figure 15.
The geothermal water flow and the thermal agent flow in relation with outdoor temperature, according to Eq. (19)-(22).

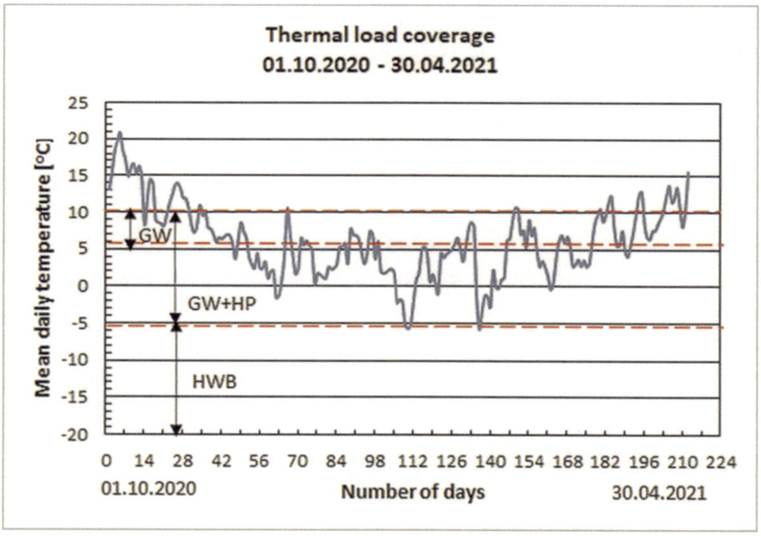

Figure 16.
The thermal load coverage in relation with outdoor temperature (source: the drawing of authors).

temperature dropped below −5°C, the temperatures in the rest of this period being in the range 0 ... + 5°C. Under these conditions, the thermal load of the system can be covered throughout the heating season without the need to come into operation the hot water boilers with gaseous fuel.

5. Conclusions

- The solution of using medium enthalpy geothermal water, from the deposits located in the lower basin of the Olt River, as a heat source for the district heating system of Călimănești is part of the current concerns to reduce the consumption of conventional fuels, generated both by depletion of reserves and the need to reduce greenhouse gas emissions. At the same time, the geothermal energy being provided free of charge by nature, continuously and renewable, allows the creation of heating systems that provide the population with heat at very low prices.

- Concerns both to increase energy efficiency and to protect reserves, have required finding solutions that allow to obtain a maximum amount of energy from available resources. In this context, the implementation in the district heating system of Călimănești city of a heat pump is a modern and very good solution for total capitalization of the thermal potential of the geothermal water available from drilling 1009. Firstly, the heat pump allows the full use of the thermal potential of geothermal water, which can be discharged at a temperature close to the environment, and secondly, increases the flow of heat sent to the district heating system, allowing either its expansion or the connection of new consumers to the network.

- The available flow of geothermal drilling water can continuously provide the necessary thermal energy for the preparation of domestic hot water. The geothermal water flow required for this purpose represents about 25% of the available well flow. During the period when the district heating system is not working, the thermal energy necessary for the preparation of domestic water can be provided entirely from geothermal water.

- During the cold season, when the district heating system comes into operation, the maximum available geothermal water flow, can only provide the thermal energy needed to prepare the domestic water and about 40% of the thermal energy needed to cover the maximum heating load of the system. Under these conditions, the district heating system can ensure the thermal comfort of consumers only up to an outside temperature of around +5°C. If the outside temperature drops below this limit, it is necessary to start the gas-fired hot water boilers. The implementation of a heat pump, which recovers the thermal potential of the wastewater discharged from the heat exchangers of the geothermal station, allows to increase the capacity of the system by about 60%. Under these conditions, given the mild climate in the area in recent years, this solution would eliminate the need to use hot water boilers with gaseous fuel, which are kept covering peak loads or possible damage.

- The water extracted from the aquifer of the area contains a large quantity of combustible gases, with a content of over 90% methane. This feature is favorable for the implementation of a gas cogeneration system, which would also provide the electricity needed for the operation of the heat pump.

Author details

Laurentiu Constantin Lipan[1*] and Sorin Dimitriu[2]

1 Faculty of Power Engineering, Department of Electric Power Systems Engineering, University Politehnica of Bucharest, Romania

2 Department of Engineering Thermodynamics, Internal Combustion Engines, Thermal and Refrigerating Equipment, University Politehnica of Bucharest, Romania

*Address all correspondence to: laurentiu.lipan@upb.ro

References

[1] U.S. Department of Energy, Geothermal Technologies Office, A history of Geothermal Energy in America [Internet]. 2022. Available from: https://www.energy.gov/eere/geothermal/history-geothermal-energy-america

[2] Britannica Encyclopedia: Geothermal Energy-Physics/History [Internet]. 2022. Available from: https://www.britannica.com/science/geothermal-energy

[3] Lund JW. Direct Heat Utilization of Geothermal Energy. In: Comprehensive Renewable Energy. Second ed. Vol. 7. Amsterdam: Elsevier; 2022. pp. 171-188. DOI: 10.1016/B978-0-12-819727-1.00119-9

[4] Toth AN. Geothermal Utilization Statistics 2019. In: Comprehensive Renewable Energy. Second ed. Vol. 7. Amsterdam: Elsevier; 2022. pp. 220-234. DOI: 10.1016/B978-0-12-819727-1.00119-9

[5] Dincer I, Ozturk M. Geothermal energy utilization. In: Dincer I, Ozturk M. Geothermal Energy Systems. Amsterdam: Elsevier, 2021, 85-136. Available from: DOI: 10.1016/B978-0-12-820775-8.00003-9

[6] Kosray M. Recent Progress in district heating with emphasis on low-temperature systems. In: Basel I, editor. Geothermal Energy, Ch. 4. London: IntehOpen; 2022

[7] Bianchi AM, Marinescu M, Hera D, Dimitriu S, Ivan G, Băltărețu F. Thermal power supply systems in Romania; development directions. In: Proceedings of the XVII-th International Conference of Engineering Thermodynamics (CNT-17); 21-22 May 2009. Brașov. Romania: Transylvania University; 2009. pp. 1-20

[8] Lund WJ, Lienau JP. Geothermal District Heating. In: Proceeding of International Geothermal Days–Conference & Summer School. 26-29 May 2009. Časta-Papiernička; Slovakia: Slovgeotherm; Session 2; Paper II-1; 2009

[9] Werner S. International review of district heating and cooling. In: Energy. Vol. 137. Amsterdam: Elsevier; 2017. pp. 617-631

[10] Rosca M, Antics M. Current status of geothermal energy utilization in Romania. In: Proceeding of International Geothermal Days – Conference & Summer School; 13-17 September 2004. Zakopane; Poland; Paper 4; 2004. pp. 295-304

[11] Bendea C, Bendea G, Rosca M, Cucuteanu D. Current status of geothermal energy utilization in Romania. In: University of Oradea, International Journal of Sustainable Energy. 2013;**4**(4):182-190. ISSN 2067-5534

[12] Dimitriu S, Bianchi AM, Baltărețu F. The up-to-date heat pump – Combined heat and power solution for the complete utilization of the low enthalpy geothermal water potential. International Journal of Energy and Environmental Engineering (IJEE). 2017;**8**(3):189-196

[13] ANRE (National Energy Regulatory Authority). Report on the state of the public service of thermal energy supply in centralized system for 2021 [Internet]. Annexes 2,3. Available from: https://www.anre.ro/ro/energie-electrica/legislatie/serviciul-public-de-alimentare-cu-energie-termica

[14] ASHRE. ANSI/ASHRE Standard 34-2019. Designation and Safety

Classification of Refrigerants. Available from: https://www.ashre.org/technical-resources/standards-and-guidelines

[15] HONEYWELL®. Products/Refrigerants. [Internet]. Available from: Refrigerants (honeywell.com)

[16] A-GAS®. Products/Refrigerants. [Internet]. Available from: Refrigerants (agas.com)

[17] MTCT (The Ministry of Transport, Construction and Tourism) C-107 Normative Regarding the Thermotechnical Calculation in Constructions. 1st ed. Romania: Best Publishing; 2008. ISBN 978-973-738-331-0

[18] Weather archive in Râmnicu Vâlcea [Internet]. Available from: https://rp5.ru/Vremea_în_Râmnicu_Vâlcea

Chapter 12

Design of Earth Quake Responses Decentralized Controller in Smart Building Systems

Doghmane Mohamed Zinelabidine and Eladj Said

Abstract

Many building systems are known to have complex structure with large dimension variables that characterize its mathematical models. In such case, it is basically desirable to use avoid the use of centralized controller due to the possibility of dimension increase during its implementation. The design of decentralized controller has faced tremendous success especially for large scale systems. The main objective of this book chapter is to design decentralized controller for building system in order to avoid the damages that will be caused by the earth quake responses. This controller is designed to increase the robustness and improve the smart building system responses toward different earth quakes. The optimized behavior of the control system has been analyzed and tested in the framework of the inclusion-contraction of the overlapping decomposition theories. Moreover, the application of this control strategy to smart building system has led to significantly minimize the damages that can be generally caused by the severe earth quakes. Thence, the obtained results have demonstrated the usefulness of the proposed controller for constructing smart cities.

Keywords: decentralized controller, earth quake responses, overlapping decomposition, smart building systems, smart cities

1. Introduction

Describing the behavior of many mechanical and engineering systems may let us end up with high dimensional mathematical models. The analysis and design problems of such systems become very complex; since the solution may not be found easily due to the huge amount of computation efforts required to simulate and analyze the dynamic process of the system, which may lead to large scale decentralized controller [1]. Therefore, new techniques and strategies should be designed in order to optimize the controller through decomposing it into simpler subsystems, so that, the control of such systems can be combined together in order to control the original global system. The objective of this paper is to design a decentralized optimal controller of well-known example of overlapping system using extension principle; our work is development of (*L. Bakule* and *J. Rodellar 1995*) paper by improving the performance of responses using optimization technique. We will see in this book chapter the application of decentralized optimal overlapping decomposition for six floor building system, for which we have given brief mathematical description of the system and the process being applied. It has been found that designing an algorithm for such type of systems is possible and very useful because it satisfies condition for this algorithm (condition of expansion/contraction, condition of contractibility of controllers). The chapter is organized as follows, in the second section, a six-floor building system mathematical model has been carefully described in order to permit the readers understand the dynamic behavior of the system. In section three, the theories of expansion/contraction have been introduced so that it will allows us to design an overlapping decentralized controller for the system under study. In section four, the contractibility of the designed controller has been discussed for application to the original system after decomposition. In the fifth section, we have proposed a controller based on the introduced theories in order to minimize the six floor building system under sever earthquake input signals. The controller robustness and performance have been demonstrated through simulation results in section six. The chapter has been ended up by conclusion and recommendation for implementation in smart building system which will be an important step toward smart cities.

2. System description

Construction engineering is very important term that gathers many disciplines that varies from physics, mechanics, electronics and control [1]. It is applied science, for which engineers build different structures within the scope of civil engineering, smart building systems; thus, it is scientific discipline to the design of building that defines smart cities [2]. People combined a practical knowledge of materials and construction with the mathematics and science that were then available [3].

Consider the mechanical second order building system shown in **Figures 1** and 2. The system is composed of six floor build in concrete cement and it is under continuous vibrations created by the continuous movement of the earth and earthquakes. The system's dynamic is described by first order differential equation written in the matrix form, the size of the matrix is proportional to the number of floors and earthquake sensors installed for each as well as the actuators at the level of each floor or set of floors. The actuators are designed to create a counter force synchronously to the building dynamic.

The system shown in **Figure 1** can be represented by the following mathematical model

$$\begin{cases} M\ddot{q} + D\dot{q} + Sq = Bu \\ y = Cq \\ v = V\dot{q} \end{cases} \tag{1}$$

Where:

$M : 6 \times 6Z$ is the mass matrix, symmetric, positive definite matrix.

$D : 6 \times 6Z$ is the damping matrix.

$S : 6 \times 6Z$ is the stiffness matrix.

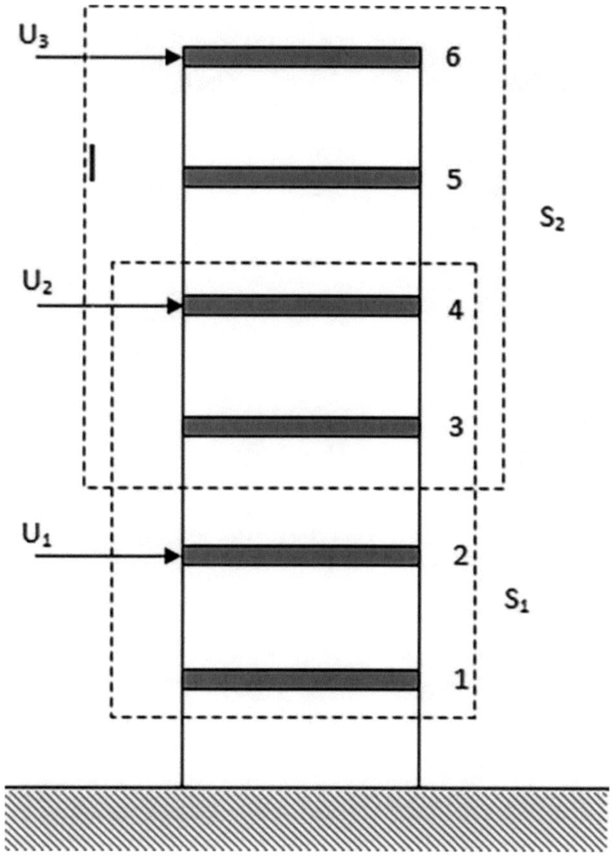

Figure 1.
Overlapping structure of building system [1].

$q : 6 \times 1Z$ is the displacement vector, represents the degree of freedom of the system.

$B : 6 \times 3Z$ is the input matrix, represents locations of actuators.

$u : 3 \times 1Z$ is the input signal, sinusoidal signal in this example.

Eq. (1) indicate the response of building system to earthquake; **Figure 3** shows a failure response to real earthquake system [1, 4].

Eq. (1) can be written as.

or

$$\begin{cases} q = T^I q_e \\ u = U u_e \\ y = G^I y_e \\ v = H^I v_e \end{cases} \tag{2}$$

Where $T^I T = I_n = I_6$, $UU^I = I_m = I_3$ $G^I G = I_p = I_6$, $H^T H = I_r = I_6$.

And T^I, U^I, G^I, H^I indicates the pseudo-inverse of T, U, G, H respectively [2].

Figure 2.
Overlapping structure of real building system [1].

Figure 3.
Failure response of a building system to earthquake disturbances [1].

$$\begin{cases} \begin{bmatrix} M_{11} & M_{12} & M_{13} \\ M_{21} & M_{22} & M_{23} \\ M_{31} & M_{32} & M_{33} \end{bmatrix} \begin{bmatrix} \ddot{q}_1 \\ \ddot{q}_2 \\ \ddot{q}_3 \end{bmatrix} + \begin{bmatrix} D_{11} & D_{12} & D_{13} \\ D_{21} & D_{22} & D_{23} \\ D_{31} & D_{32} & D_{33} \end{bmatrix} \begin{bmatrix} \dot{q}_1 \\ \dot{q}_2 \\ \dot{q}_3 \end{bmatrix} + \begin{bmatrix} S_{11} & S_{12} & S_{13} \\ S_{21} & S_{22} & S_{23} \\ S_{31} & S_{32} & S_{33} \end{bmatrix} \begin{bmatrix} q_1 \\ q_2 \\ q_3 \end{bmatrix} \\ = \begin{bmatrix} B_{11} & 0 & 0 \\ 0 & B_{22} & 0 \\ 0 & 0 & B_{33} \end{bmatrix} \begin{bmatrix} u_1 \\ u_2 \\ u_3 \end{bmatrix} \\ \begin{bmatrix} y_1 \\ y_2 \\ y_3 \end{bmatrix} = \begin{bmatrix} C_{11} & 0 & 0 \\ 0 & C_{22} & 0 \\ 0 & 0 & C_{33} \end{bmatrix} \begin{bmatrix} q_1 \\ q_2 \\ q_3 \end{bmatrix} \\ \begin{bmatrix} v_1 \\ v_2 \\ v_3 \end{bmatrix} = \begin{bmatrix} V_{11} & 0 & 0 \\ 0 & V_{22} & 0 \\ 0 & 0 & V_{33} \end{bmatrix} \begin{bmatrix} \dot{q}_1 \\ \dot{q}_2 \\ \dot{q}_3 \end{bmatrix} \end{cases} \tag{3}$$

Where dashed lines indicate the subsystems, in this example we have only two subsystems S_1 and S_2 that shares common informations $(M_{22}, D_{22}, S_{22}, B_{22}, C_{22}, V_{22})$ [1].

3. Expansion and contraction

We have the overlapping system (1) and we want to transform it into non-overlapping system described by

$$s_e : \begin{cases} M_e\ddot{q}_e + D_e\dot{q}_e + S_e q_e = B_e u_e \\ y_e = C_e q_e \\ v_e = V_e\dot{q}_e \end{cases} \tag{4}$$

To do that we consider the transformation matrices between systems (1) and (3)

$$\begin{cases} q_e = Tq \\ u_e = U^I u \\ y_e = Gy \\ v_e = Hv \end{cases} \tag{5}$$

The system described by Eq. (3) is said to be an expansion of the system described by (1) (or the system described by (1) is considered to be a contraction of the system (3)) if there exist transformation T, U, G and H that satisfied Eq. (4) so that for any initial states $(q_e(0), \dot{q}_e(0))$ and for any input $u_e(t) \in R^m$ for all $t \geq 0$ [5], we have

$$\begin{cases} q_e(0) = Tq(0) \\ \dot{q}_e(0) = T\dot{q}(0) \\ u(t) = Uu_e(t) \end{cases} \Rightarrow \begin{cases} q_e(t) = Tq(t) \\ \dot{q}_e(t) = T\dot{q}(t) \\ v_e(t) = Hv(t) \end{cases} \tag{6}$$

Essentially there exist two (02) methods to derive condition of extension:

- Method one

Requires working directly with the matrix second order equation in both original and extended system which means we need to use the matrices M and M_e [1].

- Method two

Starts by transforming the second order system into an equivalent first order system $\Rightarrow$ requires working with M^{-1} and $M_e{}^{-1}$.

Consider the system (1) and its expansion (3); define the state vectors x, x_e as: $x = \left(q^T, \dot{q}^T\right)^T$, $x_e = \left(q_e{}^T, \dot{q}_e{}^T\right)^T$ then Eqs. (1) and (2) can be written as

$$s_x : \begin{cases} \dot{x} = A_x x + B_x u \\ y_x = C_x x \end{cases} \tag{7}$$

$$s_{ex} : \begin{cases} \dot{x}_e = A_{ex} x_e + B_{ex} u_e \\ y_{ex} = C_{ex} x_e \end{cases} \tag{8}$$

Where

$$\begin{cases} A_x = \begin{bmatrix} 0_{6\times 6} & I_6 \\ -M^{-1}S & -M^{-1}D \end{bmatrix} \\ B_x = \begin{bmatrix} 0_{6\times 3} \\ M^{-1}B \end{bmatrix} \\ C_x = diag(C, V) \end{cases}$$

And

$$\begin{cases} A_{ex} = \begin{bmatrix} 0_{8\times 8} & I_8 \\ -M_e{}^{-1}S_e & -M_e{}^{-1}D_e \end{bmatrix} \\ B_{ex} = \begin{bmatrix} 0_{8\times 4} \\ M_e{}^{-1}B_e \end{bmatrix} \\ C_{ex} = diag(C_e, V_e) \end{cases}$$

Consider the transformation T, U, G satisfying (6) for the original system; define the transform for the expanded system as.

$T_d = diag(T, T)$; $C_d = diag(G, H)$.

This implies that

$$\begin{cases} x_e(0) = T_d x(0) \\ u(t) = U u_e(t) \end{cases} \Rightarrow \begin{cases} x_e(t) = T_d x(t) \\ y_{ex}(t) = C_d x(t) \end{cases} \tag{9}$$

A. Theorem one

The system s_e is an extension of the system s or equivalently s is dis-extension of s_e if and only if there exists full rank transformation matrices T, U, G and H such that

$$\begin{cases} M_e^{-1} S_e T = T M^{-1} S \\ M_e^{-1} D_e T = T M^{-1} D \\ M_e^{-1} B_e = T M^{-1} B U \\ GC = C_e T \\ HV = V_e T \end{cases} \tag{10}$$

These equations are found by transforming the system (1) and (3) into state space model [6].

Eq. (10) can be written as

$$\begin{cases} M_e^{-1} = T M^{-1} T^I + M_{cq} \\ S_e = T S T^I + S_{cq} \\ D_e = T D T^I + D_{cq} \\ B_e = T B U + B_{cq} \\ C_e = G C T^I + C_c \\ V_e = H V T^I + V_c \end{cases} \tag{11}$$

With M_e, S_e, D_e, B_e, C_e and V_e are given by

$$M_e = \begin{bmatrix} M_{11} & M_{12} & 0 & M_{13} \\ M_{21} & M_{22} & 0 & M_{23} \\ M_{21} & 0 & M_{22} & M_{23} \\ M_{31} & 0 & M_{32} & M_{33} \end{bmatrix}, \quad D_e = \begin{bmatrix} D_{11} & D_{12} & 0 & D_{13} \\ D_{21} & D_{22} & 0 & D_{23} \\ D_{21} & 0 & D_{22} & D_{23} \\ D_{31} & 0 & D_{32} & D_{33} \end{bmatrix},$$

$$S_e = \begin{bmatrix} S_{11} & S_{12} & 0 & S_{13} \\ S_{21} & S_{22} & 0 & S_{23} \\ S_{21} & 0 & S_{22} & S_{23} \\ S_{31} & 0 & S_{32} & S_{33} \end{bmatrix}, \quad B_e = \begin{bmatrix} B_{11} & 0 & 0 & 0 \\ 0 & B_{22} & B_{22} & 0 \\ 0 & B_{22} & B_{22} & 0 \\ 0 & 0 & 0 & B_{33} \end{bmatrix},$$

$$C_e = \begin{bmatrix} C_{11} & 0 & 0 & 0 \\ 0 & C_{22} & 0 & 0 \\ 0 & 0 & C_{22} & 0 \\ 0 & 0 & 0 & C_{33} \end{bmatrix}, \quad V_e = \begin{bmatrix} V_{11} & 0 & 0 & 0 \\ 0 & V_{22} & 0 & 0 \\ 0 & 0 & V_{22} & 0 \\ 0 & 0 & 0 & V_{33} \end{bmatrix}$$

Where $M_{qc}, S_{qc}, D_{qc}, B_{qc}, C_{qc}$ and V_{qc} are complementary matrices defined in such way to satisfy the condition of extension [6, 7].

A. Theorem two

The system (2) is an expansion of the system (1) if

$$\begin{cases} M_{qc}T = 0 \\ K_{qc}T = 0 \\ D_{qc}T = 0 \\ B_{qc} = 0 \\ C_{qc}T = 0 \\ V_{qc}T = 0 \end{cases} \tag{12}$$

Eq. (12) is satisfied by choosing the complementary matrices as

$$[.]_{qc} = \begin{bmatrix} 0 & 0.5[.]_{12} & -0.5[.]_{12} & 0 \\ 0 & 0.5[.]_{22} & -0.5[.]_{22} & 0 \\ 0 & -0.5[.]_{22} & 0.5[.]_{22} & 0 \\ 0 & -0.5[.]_{32} & 0.5[.]_{32} & 0 \end{bmatrix} \tag{13}$$

4. Contractibility of controllers

Let us consider the controller given by Eq. (14), for the overlapping building system

$$u = Fy + Lv + w \tag{14}$$

And Let us consider the controller given by Eq. (15) for the expanded system of the building system:

$$u_e = F_e y_e + L_e v_e + w_e \tag{15}$$

Where w and w_e are the external inputs of the smart building system [1, 8].

A. Theorem three

The controller described by Eq. (15) is contractible to the controller given by Eq. (14) if and only if

$$\begin{cases} FC = UF_eGC \\ LV = UL_eHV \end{cases} \tag{16}$$

B. Theorem four

If Eq. (2) is an extension of Eq. (1) and if Eq. (2) is stable (respectively asymptotically stable) then Eq. (1) is stable (respectively asymptotically stable) [6, 9].

5. Decentralized optimal output feedback control

A. Problem's frame

Consider the system (7); our goal is to find control law $u = Ky_x$ to minimize the cost function

$$J = \int_{-\infty}^{+\infty} \left(x^T Qx + u^T Ru\right)dt \tag{17}$$

Such that the closed loop system

$$S_c = \begin{cases} \dot{x} = (A + B_x KC_x)x \\ y_x = C_x x \end{cases} \tag{18}$$

is asymptotically stable

B. Problem Solution

First we have the following two subsystems that have been extracted from the expanded system

$$\begin{cases} M_1 \ddot{q}_{e1} + D_1 \dot{q}_{e1} + S_1 q_{e1} = B_1 u_{e1} \\ y_{e1} = C_1 q_{e1} \\ v_{e1} = V_1 \dot{q}_{e1} \end{cases} \tag{19}$$

$$\begin{cases} M_2 \ddot{q}_{e2} + D_2 \dot{q}_{e2} + S_2 q_{e2} = B_2 u_{e2} \\ y_{e2} = C_2 q_{e2} \\ v_{e2} = V_2 \dot{q}_{e2} \end{cases} \tag{20}$$

Let us transform these two subsystems into state space form to get

$$s_1 : \begin{cases} \dot{x}_{e1} = A_{ex1} x_{e1} + B_{ex1} u_{e1} \\ y_1 = C_{ex1} x_{e1} \end{cases} \tag{21}$$

$$s_2 : \begin{cases} \dot{x}_{e2} = A_{ex2} x_{e2} + B_{ex2} u_{e2} \\ y_2 = C_{ex2} x_{e2} \end{cases} \tag{22}$$

Where:

$$\begin{cases} A_{ex1} = \begin{bmatrix} 0_{(2+2)\times(2+2)} & I_{(2+2)\times(2+2)} \\ -M_1^{-1} S_1 & -M_1^{-1} D_1 \end{bmatrix} \\ B_{ex1} = \begin{bmatrix} 0_{(2+2)\times(2+2)} \\ -M_1^{-1} B_1 \end{bmatrix} \\ C_{ex1} = diag(C_1, V_1) \end{cases}$$

$$\begin{cases} A_{ex2} = \begin{bmatrix} 0_{(2+2)\times(2+2)} & I_{(2+2)\times(2+2)} \\ -M_2^{-1}S_2 & -M_2^{-1}D_2 \end{bmatrix} \\ B_{ex2} = \begin{bmatrix} 0_{(2+2)\times(2+2)} \\ -M_2^{-1}B_2 \end{bmatrix} \\ C_{ex2} = diag(C_2, V_2) \end{cases}$$

We will try to generate the optimal output feedback for each subsystem as $u_i = K_i y_i; i = 1, 2$, to each subsystem we associate the performance index:

$$J_i = \int_{-\infty}^{+\infty} (x_i{}^T Q_i x_i + u_i{}^T R_i u_i) dt \; i = 1, 2 \tag{23}$$

The necessary and sufficient conditions of optimality for each subsystem are:

$$\begin{cases} \phi_i^T P_i + P_i \phi_i + Q_i + C_{xi}^T K_i^T R_i K_i C_{xi} = 0 \\ K_i = -R_i^{-1} B_{xi}^T P_i L_i C_{xi}^T \left(C_{xi} L_i C_{xi}^T\right)^{-1} \\ \phi_i L_i + L_i \phi_i^T + X_{0i} = 0 \end{cases} \tag{24}$$

Where $\phi_i = A_i + B_{xi} K_i C_{xi}$, $X_{0i} = x_{0i} x_{0i}^T$; generally we take $x_{0i} = I$.
The optimal cost can be found as:

$$J_i = .5^* trace(P_i X_{0i}) \tag{25}$$

And the optimal control law as $u_i = K_i y_i$ where

$$K_i = \begin{bmatrix} K_{11}^i & K_{12}^i & K_{13}^i & K_{14}^i \\ K_{21}^i & K_{22}^i & K_{23}^i & K_{24}^i \end{bmatrix} \tag{26}$$

The control law for expanded system is

$$K_i = \begin{bmatrix} K_{11}^1 & K_{12}^1 & 0 & 0 & K_{13}^1 & K_{14}^1 & 0 & 0 \\ K_{21}^1 & K_{22}^1 & 0 & 0 & K_{23}^1 & K_{24}^1 & 0 & 0 \\ 0 & 0 & K_{11}^2 & K_{12}^2 & 0 & 0 & K_{13}^2 & K_{14}^2 \\ 0 & 0 & K_{21}^2 & K_{22}^2 & 0 & 0 & K_{23}^2 & K_{24}^2 \end{bmatrix} \tag{27}$$

and the contracted control law for the original system is:

$$K = \begin{bmatrix} K_{11}^1 & K_{12}^1 & 0 & K_{13}^1 & K_{14}^1 & 0 \\ K_{21}^1 & K_{22}^1 + K_{11}^2 & K_{12}^2 & K_{21}^1 & K_{24}^1 + K_{13}^2 & K_{14}^2 \\ 0 & K_{21}^2 & K_{22}^2 & 0 & K_{23}^2 & K_{24}^2 \end{bmatrix} \tag{28}$$

To apply this control law for the original mechanical system, we must write it in the form:

$$u = Fy + Lv + w \tag{29}$$

Where: w is the external input to the mechanical system [6, 10].
We have

$$\begin{cases} K = [F, L] \\ K_e = [F_e, L_e] \end{cases} \tag{30}$$

With

$$\begin{cases} F = \begin{bmatrix} K_{11}^1 & K_{12}^1 & 0 \\ K_{21}^1 & K_{22}^1 + K_{11}^2 & K_{12}^2 \\ 0 & K_{21}^2 & K_{22}^2 \end{bmatrix} \\ L = \begin{bmatrix} K_{13}^1 & K_{14}^1 & 0 \\ K_{23}^1 & K_{24}^1 + K_{13}^2 & K_{14}^2 \\ 0 & K_{23}^2 & K_{24}^2 \end{bmatrix} \end{cases} \tag{31}$$

Now the projection of this control law onto the original mechanical system gives [11]:

$$\begin{cases} M\ddot{q} + (D + BLV)\dot{q} + (K + BFC)q = Bw \\ y = Cq \\ v = V\dot{q} \end{cases} \tag{32}$$

6. Simulation results and discussion

Results founded in this paper using Matlab environment, where first, we started with centralized non-optimal controller for each floor as shown in figures below (**Figures 4–6**).

Then we apply decentralized non-optimal controller for the same floors (**Figures 7–10**).

We notice that in common floor (**Figures 8** and **10**) the response of system in closed loop form still effective; this is according to the interconnection between subsystems **Figure 11**.

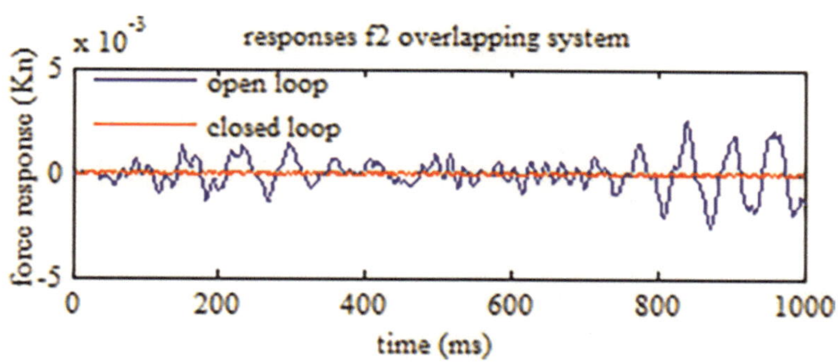

Figure 4.
Centralized non-optimal control system (flour 2).

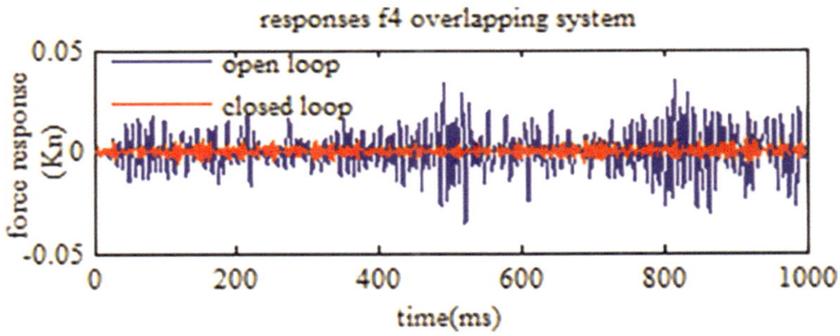

Figure 5.
Centralized non-optimal control system (flour 4).

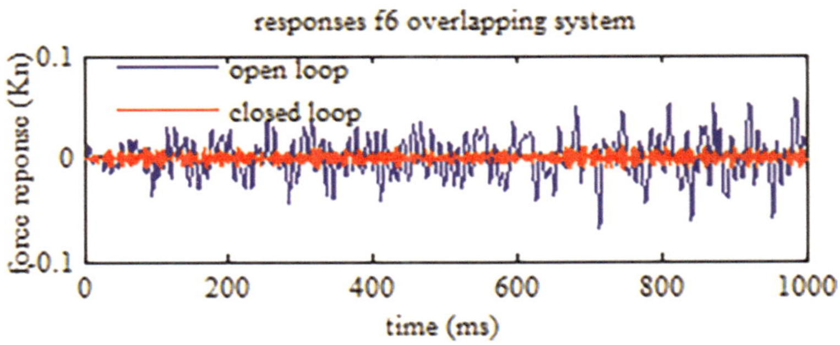

Figure 6.
Centralized non-optimal control system (flour 6).

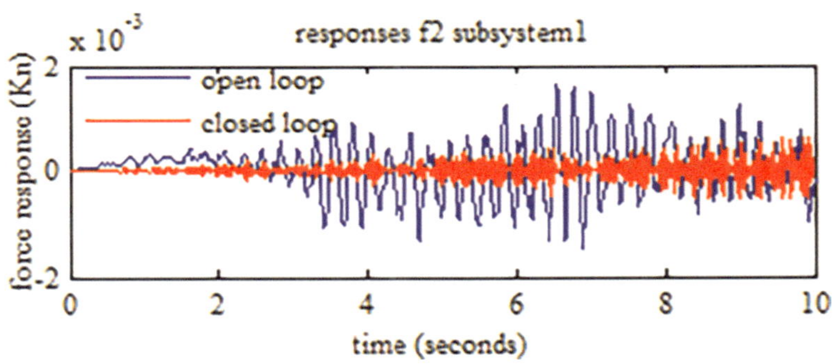

Figure 7.
Decentralized non-optimal control sub-system 1 (flour 2).

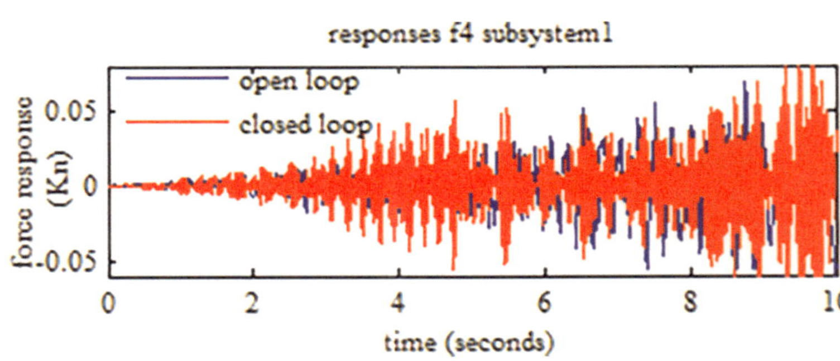

Figure 8.
Decentralized non-optimal control sub-system 1 (flour 4).

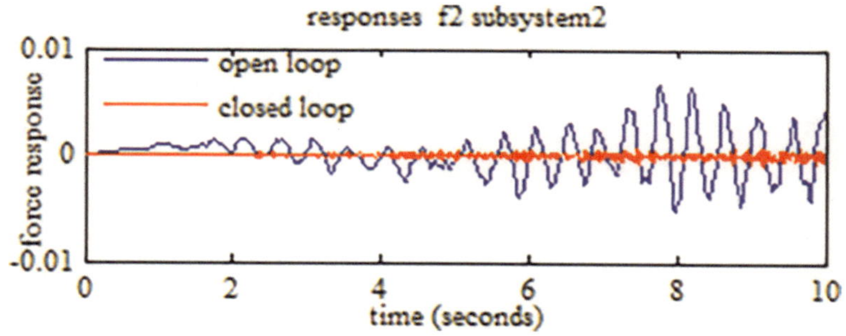

Figure 9.
Decentralized non-optimal control sub-system 2 (flour 2).

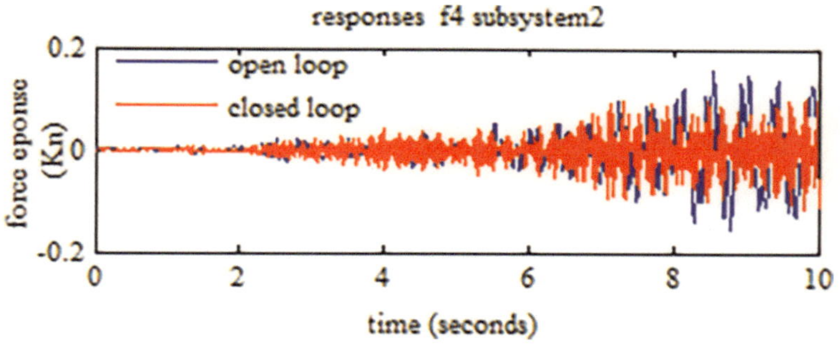

Figure 10.
Decentralized non-optimal control sub-system 2 (flour 4).

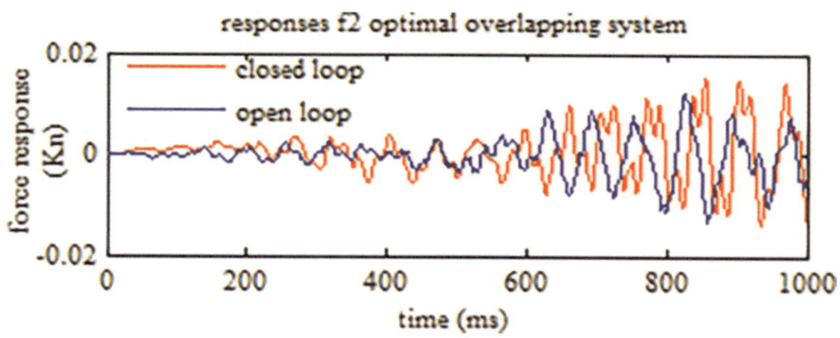

Figure 11.
Centralized optimal control system (flour 2).

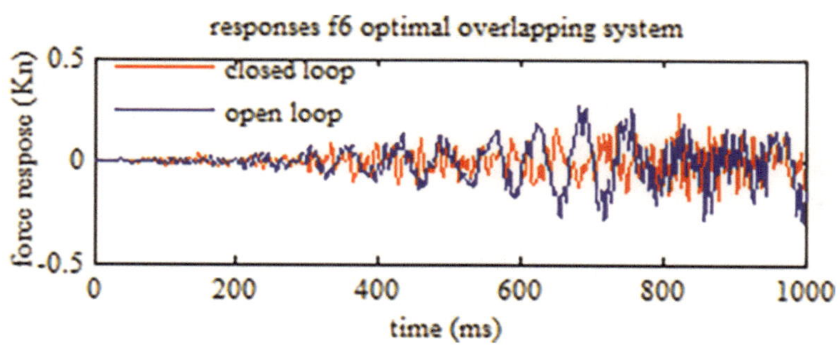

Figure 12.
Centralized optimal control system (flour 6).

An optimization technique was applied for centralized and decentralized controller as show in figure below:

Even when we applied optimization responses in **Figure 12** still valuable which may make damages to the building. The cost function of centralized controller is $J_t = 7.31^* 10^3$ while decentralized controller gives (**Figures 13–16**).

The cost functions of subsystems 1 and 2 are respectively $J_1 = 2.47^* 10^3$ and $J_2 = 1.41^* 10^3$.

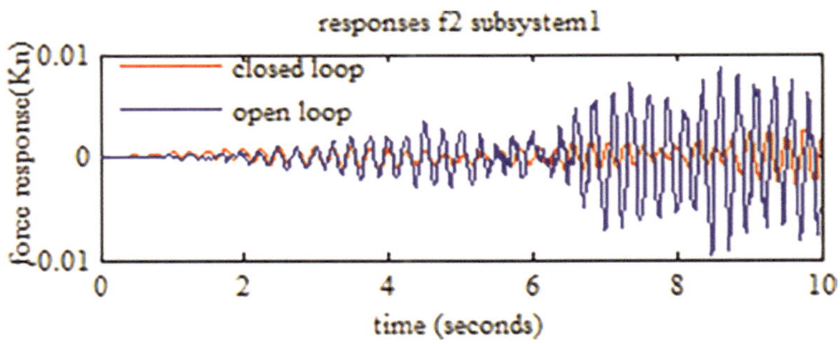

Figure 13.
Decentralized optimal control sub-system 1 (flour 2).

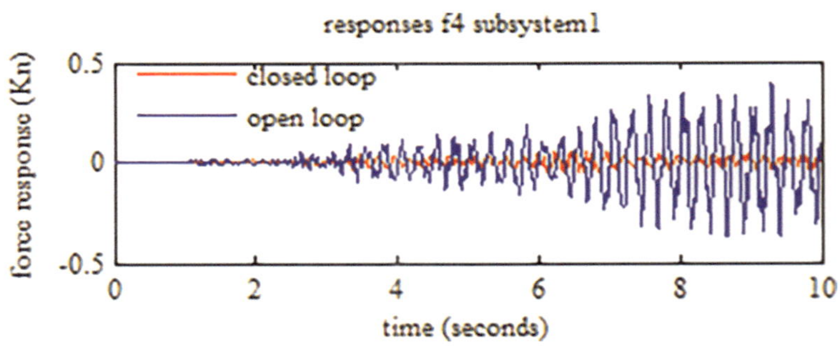

Figure 14.
Decentralized optimal control sub-system 1 (flour 4).

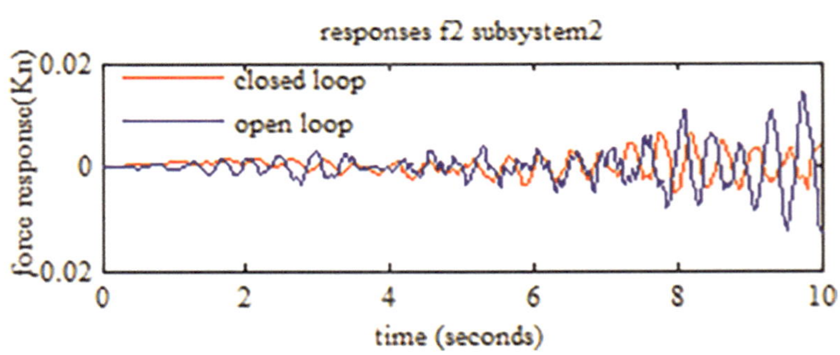

Figure 15.
Decentralized optimal control sub-system 2 (flour 2).

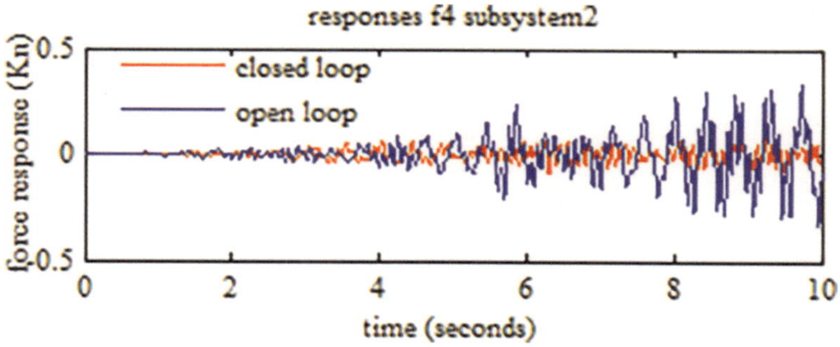

Figure 16.
Decentralized optimal control sub-system 2 (flour 4).

7. Conclusion

The overlapping decomposition method has been presented in this book chapter, and then the inclusion principle has been introduced to provide a mathematical framework for decentralized control. The inclusion of the cost function has also been discussed to incorporate the optimal control problem for large scale smart building systems where the concept of contractibility of controllers has been discussed. Optimal decentralized dynamic output feedback controllers design has been proposed for six-floor building systems with overlapping structure. Non-optimal overlapping centralized and decentralized controllers are designed for which we found that decentralized controller give better results. Furthermore to improve these results we developed an optimization technique that allow us not just design optimal controller but also minimize the cost function of the whole system for decomposed decentralized ($J_1 = 2.47^* 10^3, J_2 = 1.41^* 10^3$) in comparing to centralized ($J_t = 7.31^* 10^3$) controller; where we it is clear that $J_1 + J_2 \prec J_t$. The obtained results have demonstrated the superiority of the proposed controller to design smart building system toward smart cities.

Acknowledgements

This paper is a summary of final year project' part for getting research master of M. Z. Doghmane, at the department of electrification (Ex-INH), university of Boumerdes, Algeria.

Author details

Doghmane Mohamed Zinelabidine[1]* and Eladj Said[2]

1 Faculty of Hydrocarbons and Chemistry, University M'hamed Bouguerra, Algeria

2 Department of Automation and Electrification of Industrial Procedures, Algeria

*Address all correspondence to: doghmane_m@yahoo.com

References

[1] Doghmane MZ, Kidouche M, Eladj S, Belahcene B. Design of Optimal Decentralized Controller Using Overlapping Decomposition for smart building system. In: Hatti M, editor. Artificial Intelligence and Renewables Towards an Energy Transition. ICAIRES 2020. Lecture Notes in Networks and Systems. Vol. 174. Tipasa, Algeria: Springer, Cham; 2021. DOI: 10.1007/978-3-030-63846-7_16

[2] Mendil C, Kidouche M, Doghmane MZ. Modeling of hydrocarbons rotary drilling systems under torsional vibrations: A survey. In: Hatti M, editor. Artificial Intelligence and Renewables Towards an Energy Transition. ICAIRES 2020. Lecture Notes in Networks and Systems. Vol. 174. Tipasa, Algeria: Springer, Cham; 2021. DOI: 10.1007/978-3-030-63846-7_24

[3] Reitherman R, Anagnos T, Meluch W. Building Bridges between Civil Engineers and Science Museums. USA: Consortium of Universities for Research in Earthquake Engineering; 2008

[4] Schweier C, Markus M, Steinle E. Simulation of earthquake caused building damages for the development of fast reconnaissance techniques. Natural Hazards and Earth System Sciences. 2004;**2004**(4):285-293

[5] Iftar L. Decentralized estimation and control with overlapping input-state output decomposition. Automatica. 1993;**29**:511-516

[6] Doghmane MZ, Kidouche M, Ahriche A. Decentralized overlapping control design with application to rotary drilling system. IETE Journal of Research. 2021. DOI: 10.1080/03772063.2021.1886602

[7] Bakule L, Lunze J. Decentralized Design of Feedback Control for Large-Scale Systems. Priloha casopisu Kybernetika, Supplement to the journal Kyberneitika: ACADEMIA, praha; 1988

[8] Mendil C, Kidouche M, Doghmane MZ. A study of the parametric variations influences on stick-slip vibrations in smart rotary drilling systems. In: Hatti M, editor. Artificial Intelligence and Renewables Towards an Energy Transition. ICAIRES 2020. Lecture Notes in Networks and Systems. Vol. 174. Tipasa, Algeria: Springer, Cham; 2021. DOI: 10.1007/978-3-030-63846-7_67

[9] Ikeda M, Siljack DD. Overlapping decentralized control with input, state, and output inclusion. Control Theory and Advanced Technology. 1986;**2**(2):155-172

[10] Palacios-Quinonero F, Rodellar J, Rosell JM. Sequential design of multi-overlapping controllers for longitudinal multi-overlapping systems. Applied Mathematics and Computation. 2010;**217**(3):1170-1183

[11] Mendil C, Kidouche M, Doghmane MZ. Automatic control of a heat exchanger in a nuclear power station: The classical and the fuzzy methods. International Conference on Advanced Electrical Engineering (ICAEE). 2019;**2019**:1-6. DOI: 10.1109/ICAEE47123.2019.9014661

Chapter 13

The State of Renewable Energy in China and Way Forward in New Scenario Policies

Daniel Adu, Ransford O. Darko, Boamah Kofi Baah and Agnes Abbey

Abstract

China has a huge undeveloped potential in the field of renewable energy and represents the highest nation in hydropower both in installation and generation worldwide with which contributes greatly in their energy supply and helps in elimination energy demand situation in the country to direct the energy system toward the area of sustainably, clean, and quality energy supply for its safety in the country. The manuscript discusses the current renewable energy systems in China, the potentials for various renewable energy sources, and way forward. It also discusses the expected state of renewable energy in the next four decades in China and the various strategic development techniques. The current renewable energy situations, both demand and supply situations, have been discussed. Summary of the approaches for new development by the various energy sectors in the country as well as support of the government in renewable energies has been analyzed.

Keywords: renewable energy, potential, energy consumption, hydropower, wind power, solar power

1. Introduction

With the advent of renewable energy and high investment risk, the main financial support areas of R & D activities for commercial applications are REPG [1], among the significant financial support for the Chinese government's special funds for renewable energy, for the development of renewable energy-related activities, including resource exploration, and the development of standards, as well as the development of the project [2]. A small number of specific regulations were allotted for the special fund. For example, the Operation Views on Promoting the Development of Wind Power Generation Industry in 2006 was released by NDRC and MOF. It specified that the government would select several competent wind turbine and component manufacturers in which the R&D and testing of new products could be supported by the Renewable Energy Development Special Fund [3]. In the same way, appropriate laws and regulations also specified that the R&D of new products, as well as site industrialization of key technologies in the solar power industry, will be supported by a portion of the special fund [4, 5]. In addition to the special funds for the development of renewable energy, the government of China has as well provided R&D funds for renewable energy through the following plans and programs, for example, The National Basic Research Program of China ("973" Program), The "Five-year Plans," and The National Natural Science Fund and The National High Technology Research and Development Program of China ("863" Program), [6]. Similarly, Liu et al., on the other hand, have divided China's current renewable energy policy into several categories, including development planning, industry guidance and technical support, cost sharing, price incentives, legal responsibility, and promotion [7].

2. China's energy structure (supply and demand)

China's energy demand is currently led by industrial sector, but there will be a drastic change in future. While total energy demand will remain as present in 2050, the structure will change. There will be a rise in the transport and building sectors energy consumption, while the industrial sector consumption will be reduced. Based on the policy report, the final energy demand is expected to reach 3397 Mtce by the year 2050 [8]. This rising demand can be resolved through renewable sources. This is true for both scenarios; nevertheless, the electrification and share of renewable energy are great in the Below 2°C Scenario. Here, 54% of energy demand will be electrical energy in 2050 in the scenario compared to 37% in the Stated Policies Scenario. Industrial area fossil energy use is mainly substituted by electricity. China is on the path to a greener and more diversified energy supply. The higher dependence on coal is removed and substituted with nonfossil energy sources. In both scenarios, the energy demand will peak around 2030, and by 2050 the Below 2°C Scenario will have an energy demand of 3202 Mtce [8]. **Figure 1** shows the main energy installed capacity and peak demand in China by regions. **Figure 2**(**a**, **b**) shows the final energy demand (Mtce) in 2050 in the two scenarios compared with today, by sector and fuel type. There was an increase in total energy production from 627.7 million tce to 3600 between 1978 and 2014, with the consumption also rising by 5.69 times during the same period, attainment of 4260 million tce in 2014. **Figure 3** shows China's Primary Energy Consumption between 2010 and 2016. **Figure 4** compares the energy consumption by source in China between 2010 and 2016 [8].

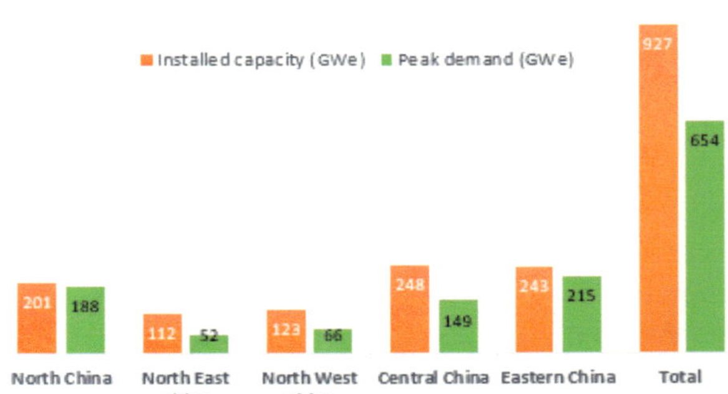

Figure 1.
The main energy installed capacity and peak demand in China by regions.

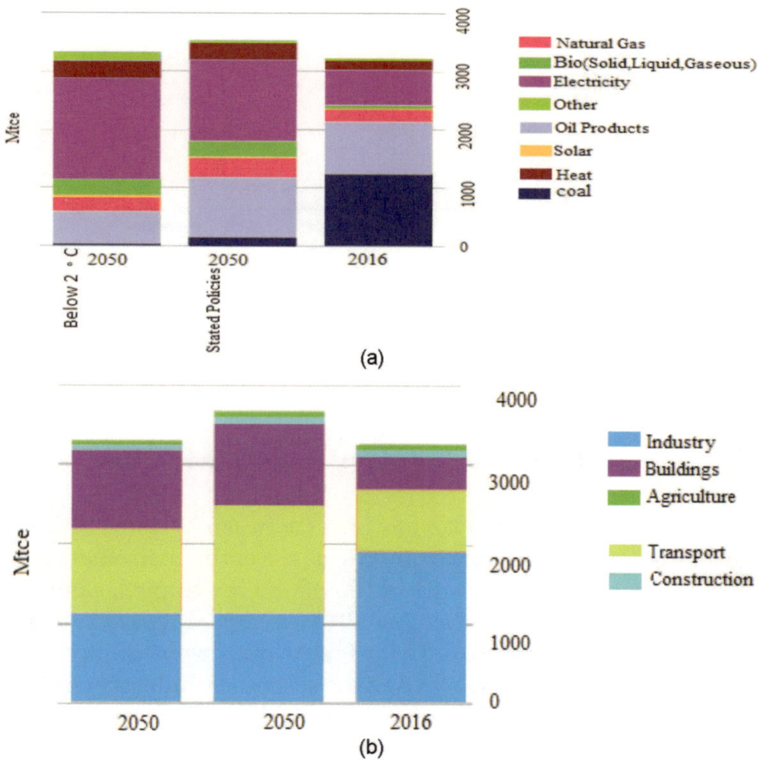

Figure 2.
Final energy demand (Mtce) in 2050 in the two scenarios compared with today, by sector and fuel type.

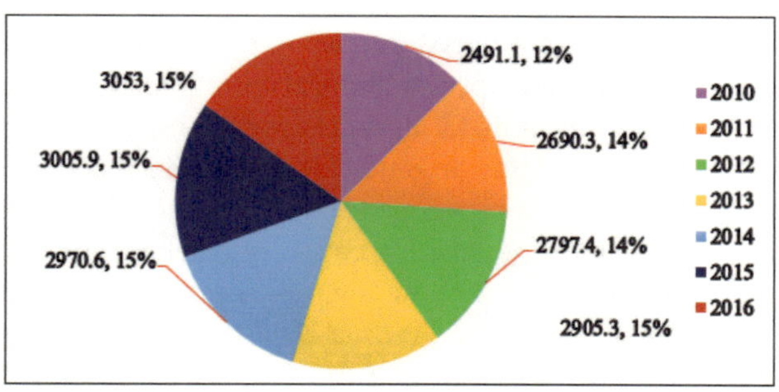

Figure 3.
*China's Primary Energy Consumption between 2010 and 2016. * Primary energy comprises commercially traded fuels, including modern renewables used to generate electricity. Notes: Oil consumption is measured in million tonnes; other fuels are measured in million tonnes of oil equivalent.*

As stated by Li Fulong, consumption of coal during the first half of 2017 was about 1.83 billion metric tons making 59.8% of the total energy. There was a rise of 3% in natural gas and nonfossil fuels mix to 29%. Based on the energy sector's 5-year plan for 2016–2020, China targets to bring down the country's share of coal energy mix down to less than 58%. China's total energy consumption is expected to reach 5 billion tons by 2020, bringing about an annual increase of approximately 2.5% between 2016 and 2020. As stated by the plan, there will be an increase in the share of nonfossil fuels to over 15% with the share of natural gas also reaching 10% [9].

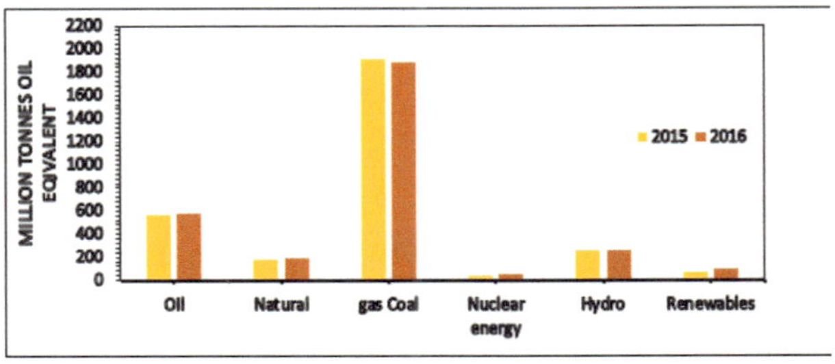

Figure 4.
*China's Primary Energy Consumption by fuel type between 2010 and 2016. * Primary energy comprises commercially traded fuels, including modern renewables used to generate electricity. Notes: Oil consumption is measured in million tonnes; other fuels are measured in million tonnes of oil equivalent.*

3. Current situation of renewable energy resources in China

As stated by the 13th Five-Year Plan of China (between 2015 and 2020), energy from nonfossil was expected to constitute 15% of the entire main energy consumption during 2020. Based on the statistics of China's energy council, China's national total electricity consumption as on October 2017 was 513.0 billion kWh, an increase of 5.0% over the same period in 2016. The primary industry consumption was 8.4 billion kWh of electricity, an increase of 3.6%; the secondary industry consumption also was 365 billion kWh of electricity, an increase of 3.0%; tertiary industry consumption was 70.8 billion kWh of electricity, an increase of 12.4%; and urban and rural resident's consumption was 68.8 billion kWh of electricity, an increase of 8.7%. All increased during the same time in 2016. The national installed capacity also reached 1.67 billion kW, and thus, 6000 kW and over power plants, an increase of 7.3% compared to 2016. Thus, hydropower was 300 million kW, thermal power was 1.08 billion kW, nuclear power was 35.82 million kW, and wind power was 160 million kW [10]. On the side of generation, the hydropower generation capacity reached 923.4 billion kWh within the first 10 months of 2017, an increase of 2.2% compared to the same time in 2016. Others like thermal power generation capacity reached 3799.3 billion kWh, an increase of 5.4%, nuclear power generation reached 203.6 billion kWh, an increase of 18.4, and wind power generating capacity reached 239.7 billion kilowatts, an increase of 25.3%. **Figures 3** and **4** show the power production and installed capacity in the country as at the end of October 2017 with the year-on-year change in % [11] (**Figures 5** and **6**). **Figure 7** shows Electricity Generation from Renewables in China by source, 2008–2015.

According to China's electricity council, power system in China is still subjugated by coal; nevertheless, the 2016 additional installed capacity of solar power of 77 GW and wind power and solar power total of up to 149 GW brought about the higher portion of renewable installed capacity. Wind turbines generated 241 TWh of electricity in 2016, thus, an increase of 30% on that of 2015. Solar power generation also adds up by 72% with an increase in generation of 66TWh in 2016. Nuclear power stations also saw growth in 2016 compared to 2015 with a generation of additional 24% electricity reaching 213 TWh [13]. **Figure 8** shows the installed power generating capacity in China together with the additions in 2016, while **Figure 9** shows year-on-year growth in power production as **Figure 10** shows the energy generation mix in the country 2016.

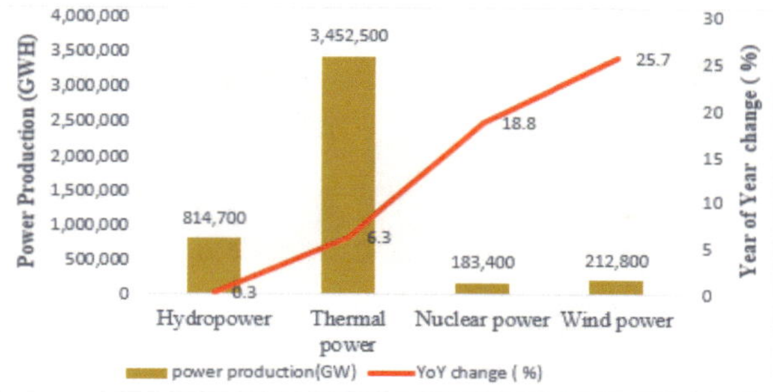

Figure 5.
*Power production from renewables in China 2017. *Thermal includes coal, gas, oil, and biomass.*

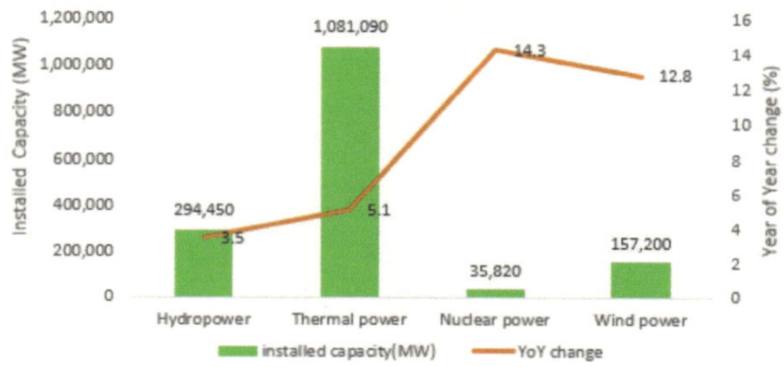

Figure 6.
*Renewable energy installed capacity in China 2017. *Thermal includes coal, gas, oil, and biomass.*

4. Renewable energy potential

China has a huge undeveloped potential in the field of renewable energy, and the government has recently been aware of these resources and begun to take steps to take advantage of that potential. A study has shown that the country will be able to meet the emerging new demand if all the targets set by the government are achieved as it will yield new renewable energy generation capacity of 362 GW [15].

4.1 Hydropower

China leads globally in the field of hydropower installment capacity and generation. The country in 2016 saw growth in its overall installed hydro capacity of about 11.74 GW making a total of 330 GW [16]. China is the leading country of hydropower development in the world with the world's leader in hydropower scale and project capacity development, with the largest hydropower station in the world's situated along the Yangtze River. China had 117 GW of installed capacity in 2005 with the aforementioned of 300 GW by 2030. According to World Watch Institute, China has hydropower potential capacity of 500 GW [17]. The strong water resources in China are mainly in the western underdeveloped and along the east coast where electricity demand increases. Most of China's water resources are in Sichuan, Tibet, and Yunnan, where the Yangtze River begin in Tibet and flows through Sichuan; the Pearl began in Yunnan [18]. China has greatly made a perfect use of the various most important channels in the country for hydroelectric development. **Figure 3** shows The Three

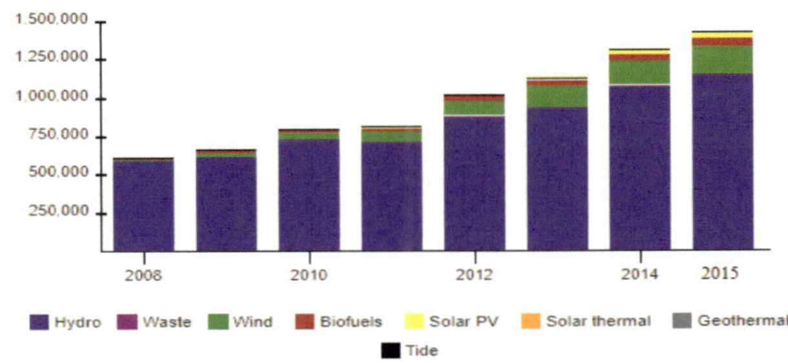

Figure 7.
Electricity Generation from Renewables in China by source, 2008–2015 (GWh) [12].

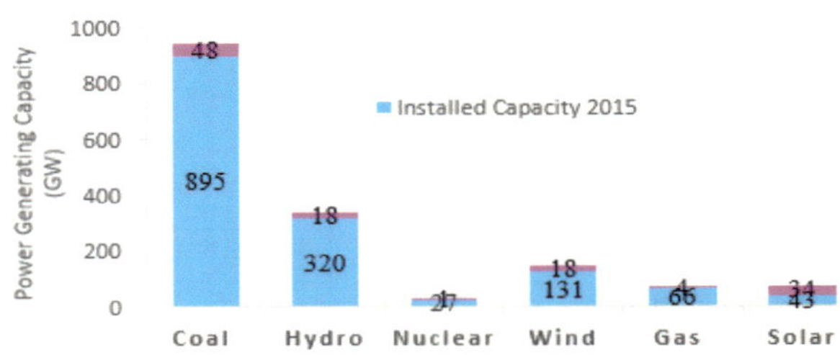

Figure 8.
The installed power generating capacity in China together with the additions in 2016 in GW [13].

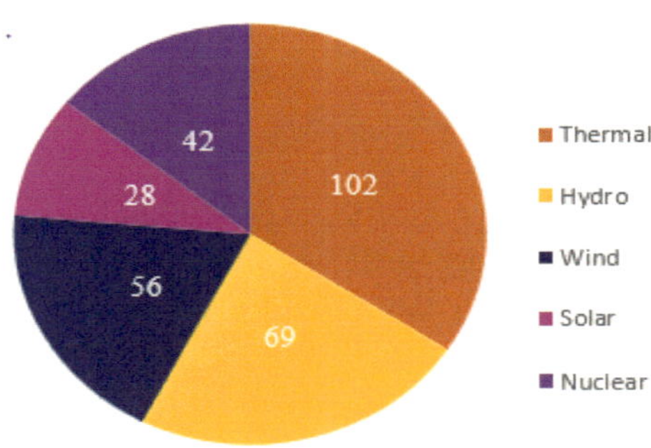

Figure 9.
Year-on-year growth in power production (TWh) [13].

Gorges Dam, while **Figure 4** shows the largest hydroelectric plants in China with their generation and installed capacity (**Figure 11**) [19].

4.2 Wind

With hydropower, wind power has a great opportunity to become a major source of renewable energy in China. The Chinese government has developed a potential power generation capacity for the development of China's wind power market, with its ability to increase its planned capacity by 2020 to generate 30 GW. China experienced a significant increase in energy from the wind farm throughout the country. [20]. The wind power industry in China has succeeded in rapid growth due to government intervention. China is a world leader in wind power generation, with the largest installed capacity of any nation [21] and continued rapid growth in new wind

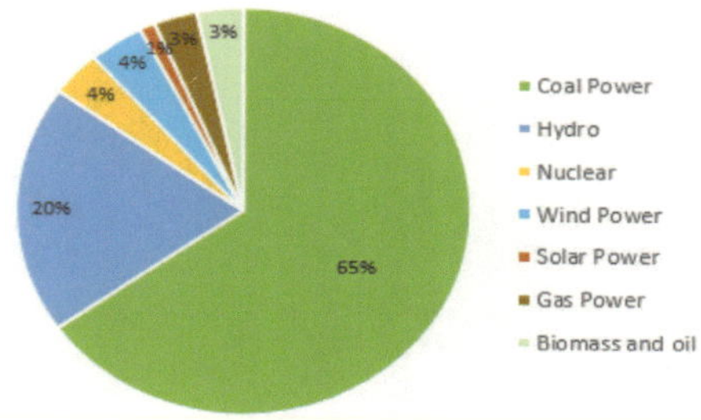

Figure 10.
China's electricity mix in 2016 in TWh [14].

Figure 11.
The Three Gorges Dam.

facilities [22]. In 2016, the country added 19.3 GW of wind power generation capacity [23] reaching a total capacity of 149 GW, [12] with an electricity generation of 241 TWh representing 4% of national total electricity consumption. China's projection is to attain 250 GW of wind capacity by 2020 as part of government's promise to produce 15% of all electricity from renewable resources [24]. The status of the base has been included in the policy and market with the development of tax incentives. In 2004, the Commission carried out a "wind power concession" for a 20-year work cycle to lessen the in-grid wind power tariff through the establishment of wind farms with high capacity [25]. Prior to 2006, the Commission agreed to construction of five large wind farms, all of having no less than 100 (MW) capacity [25]. As shown **Figure 12**. When the law of renewable energy was put into effect in 2006, the grid company was obliged to sign a grid connection agreement with the wind power generating company and purchase the full amount of the wind power generated by it [25]. **Table 1** shows wind power installed capacity (MW) and generation (GWh) in China 2007–2016.

4.3 Biomass power

Biomass is a multistep process of producing synthetic hydrocarbon fuels made from biomass that can be converted to both solid and liquid and gaseous fuels through

Small wind turbines and solar power panels on lamp posts in Minhang District, Shanghai	A wind farm in Xinjiang, China

Figure 12.
Small wind turbines and solar power panels as well as wind farm in China.

chemical and biological processes [26]. Biomass is anticipated to contribute roughly between 15% and 50% of the world's primary energy consumption by the year 2050 [27]. Biomass in China is a larger energy source than most would think because of the huge rural population. Eighty percent of biomass energy is located in rural China with the principal source being crop residue [28]. Approximately 4 billion tons of crop residues and wood fuels are burnt using stoves in the western rural areas [29] The Chinese government is making great strides to develop rural renewable energy, because biomass is a good already established one, and they are considering further development and efficiency gains in the use of traditional biomass. The government has begun to promote the development of biomass energy to achieve multiple projects at the national level. They participated in the planting and reforestation program in Wuhan, Guangdong Province. Over the past 14 years, the region has grown to 170.600 hectares with an increase in coverage from 31.5% to 49.4%, almost 20% of the increase in [29] and 62.8% increase in annual production capacity, and the increase can be repeated in other parts of the country. Due to the abundant domestic biomass resources, the biomass energy industry in China is rapidly developing [30].

4.4 Solar

China is leading globally in terms of solar PVs installation since 2013 and leading the world's largest market for both photovoltaics and solar thermal energy. It increased its total PV capacity to 77.4 GW [31] and was the first country to pass 100 GW of cumulative installed PV capacity in 2017 [32]. China currently has six factories that produce no less than 2 GW/year each of monocrystalline, polycrystalline, and noncrystalline photovoltaic cells. They comprise LDK Solar Co, Wuxi Suntech Solar Energy Co., Ltd. 50 MW/year of solar, Yunnan Semi-Conductor Parts Plant 2 MW/year of mono-crystalline cells, the Baoding Yingli Solar Energy Modules Plant, 6 MW/year of polycrystalline cells and modules, the Shanghai Jiaoda Guofei Solar Energy Battery Factory 1 MW/year of modules, and the Shanghai PV Science and Technology Co., Ltd. 5 MW/year of modules [32]. **Figure 13** shows SolarGIS-Solar-map-China-Mainlands.

Some of the technologies used as solar collectors and photovoltaic modules include the following: Molten Salt Storage Technology, the process uses inorganic

DOI: http://dx.doi.org/10.5772/intechopen.106492

	2008	2009	2010	2011	2012	2013	2014	2015	2016
Capacity (MW)	12,200	16,000	31,100	62,700	75,000	91,424	114,763	129,700	149,000
Production (GWh)	3675	5710	14,800	26,900	44,622	74,100	103,000	134,900	153,400

Table 1.
Wind power installed capacity (MW) and generation (GWh) in China 2007–2016 [24].

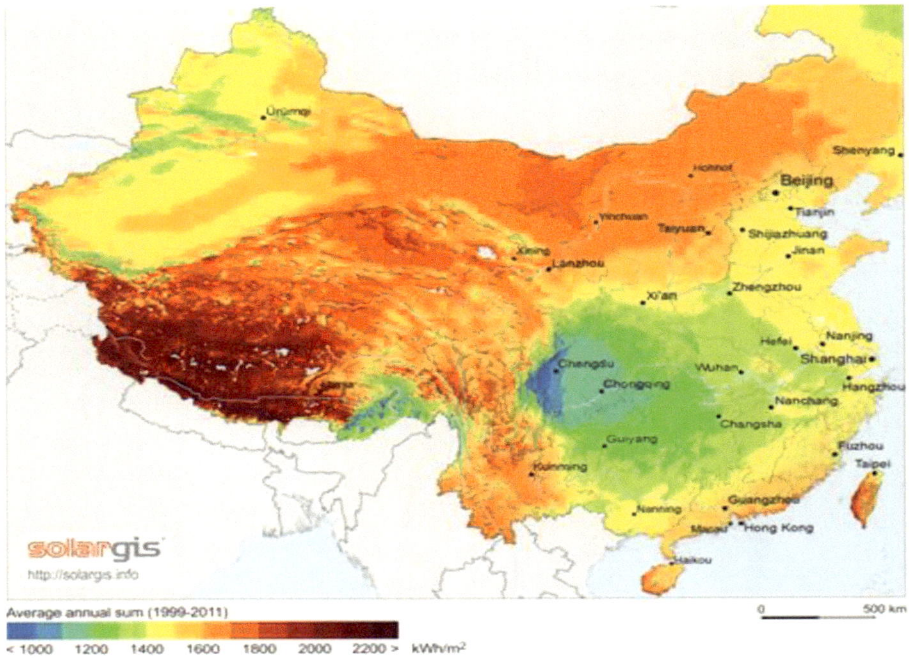

Figure 13.
SolarGIS-Solar-map-China-Mainlands.

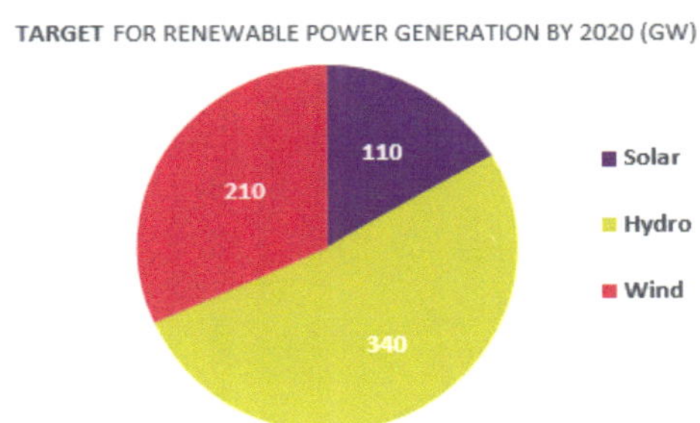

Figure 14.
Target for renewable energy generation by 2020 (GW).

salts to transfer energy generated by solar PV systems into solar thermal using heat transfer fluid rather than oils as some storage system have; and Solar Panel with Built-In Battery: with this application, the rechargeable battery is built into the solar panel itself, rather than operating as two stand-alone systems. According to scientists, conjoining the two into one system could lower costs by 25% compared to existing products.

5. Perspectives of renewable energy development in China

Studies are currently being conducted on other renewable energy sources, such as energy marine current energy, wave energy, ocean thermal energy, and salinity

DOI: http://dx.doi.org/10.5772/intechopen.106492

	Solar	Wind	Biomass	Hydro	Nuclear	Gas	Coal
2010	0	30	0	216	0	36	655
2016	77	149	0	332	34	67	945
2025	340	336	25	400	86	145	1061
2030	469	429	33	440	111	173	1089
2035	600	515	41	471	135	199	1096
2040	739	593	49	493	145	219	1087

Table 2.
China's new policies scenarioi capacity by technology (GW).

gradient energy, not forgetting tidal energy. Nevertheless, they are hardly ever being used for commercial power generation due to high cost, poor dependability, low efficacy, poor stability, and small size [33, 34]. The total accessible reserve of ocean energy resources in China is anticipated to be about 1000 GW with great potential for exploitation [35]. **Figure 14** shows the Target for Renewable Energy Generation by 2020 and **Table 2** also shows China's New Policies Scenario Installed Capacity by Technology [36].

5.1 Switching to a cleaner source of power

A good distribution and policy support helps to make renewables cheaper, and solar PVs turn out to be the cheapest form of energy generation in China.

Installed low-carbon capacity, led by hydropower, wind, and solar PV, raises quickly and constitutes 60% of total capacity by 2040. Average solar PV projects in China become cheaper than both new and existing gas-fired power plants around 2020 and cheaper than new coal-fired capacity and onshore wind by 2030 [37].

There was a rise of 1.3% in China's energy consumption in 2016. Increase during these 2 years (2015 and 2016) has been the lowest aside 1997–1998. In spite of this, China continues to be the world's largest growth market for energy within 16th successive years. Renewable power (without hydro) increased by 14.1% in 2016, beneath the 10-year average, but the largest rise on record (53 Mtce). The wind gave over half of renewables growth, while solar energy contributed almost a third despite accounting for only 18% of the total [2]. According to China's energy agency, the country will cultivate $361 billion into renewable power generation by 2020 which will create over 13 million jobs in the sector. This is containing in the National Energy Administration (NEA) blueprint document which outlines their agenda for national energy sector development throughout their 5-year 2016 to 2020 period. The NEA also mentions that renewable power installed capacity including wind, hydro, solar, and nuclear power will constitute about half of new electricity generation by 2020 [38]. **Figure 15** shows China's Primary Energy Consumption by a source in the new portfolio (**Figure 16**).

Based on IRENA the global roadmap, renewable energy portion in China's energy mix is anticipated to reach 17% by 2030 comparable to 13% in 2010. This analysis, however, displays that the country can convincingly achieve modern renewables to

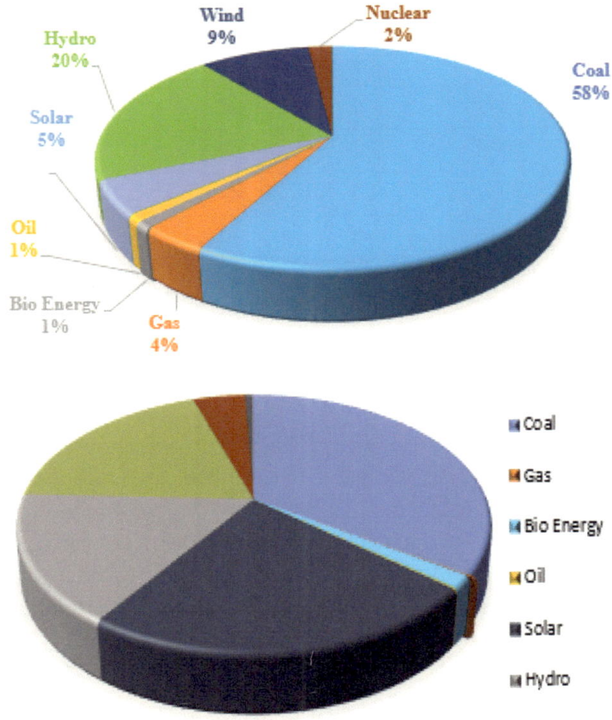

Figure 15.
Installed power generation capacity in China in the New Policies Scenario 2040 of a total of 3188GW.

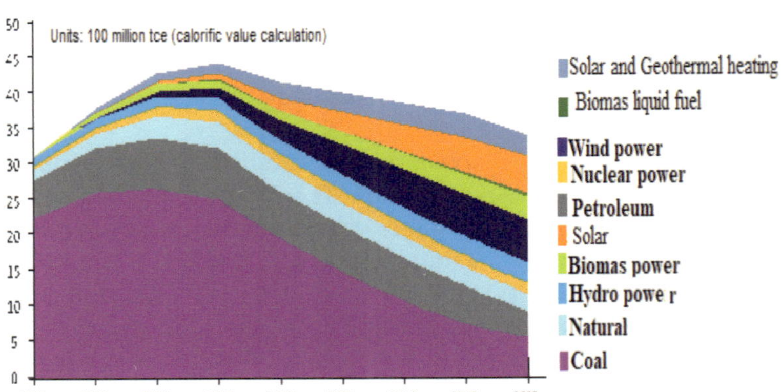

Figure 16.
China's primary energy consumption.

26%. China can save over USD 200 billion a year by 2030, helping in the improvement of health and reducing CO2 emissions [39]. Under the high renewable energy penetration scenario, total power generation in China will be 15.2 trillion kWh in 2050. This consist of 1038 kW coal, 466 kW natural gas, 649 kW nuclear, 2187 kW hydro, 5350 kW wind, and 4130 kW of solar. All these energy resources were in billions. This, therefore, constitutes total of 85.8% of renewable energy power generation and 91% of nonfossil energy [40].

Power generation capacity (GWe)	**2020**				**2030**				**2050**			
	G1	**G2**	**G3**	**G4**	**G1**	**G2**	**G3**	**G4**	**G1**	**G2**	**G3**	**G4**
Hydropower	310	340	370	350	403	430	453	400	440	510	530	530
Wind power	50	100	150	200	120	180	300	400	300	500	800	1000
Solar power	5	20	30	80	50	100	200	450	500	800	1000	1300
Geothermal power	0.08	0.08	0.08	0.5	0.5	0.5	0.5	5	1.5	1.5	1.5	50
Biomass power	15	20	30	30	20	20	20	35	10	10	10	35
Ocean power	0.04	0.04	0.04	0.1	1	1	1	1	10	10	10	10
Biogas	34.6	34.6	34.6	34.6	62.9	62.9	62.9	61	78.6	78.6	78.6	108
Biomass briquette	7.1	14.3	21.4	25	14.3	21.4	35.7	25	21.4	35.7	57.1	40
Geothermal	10.3	10.3	10.3	50	20.5	20.5	20.5	100	41	41	41	200
Ethanol/biodiesel	16.4	24.2	31.1		44.4	77.9	101.1	56.6	77.9	134.9	168.6	193
Total renewables (billion tce)	0.51	0.61	0.73	0.86	0.73	0.89	1.1	1.4	1.14	1.54	1.93	2.6
Total energy consumption (billion-tce)	4	4	4	4.6	4.5	4.5	4.5	5.5	5	5	5	5.8
Share of renewables (%)	12.7	15.3	18.2	18	16.3	19.8	24.4	25	22.8	30.8	38.5	45

a G1-G3 are quoted from [15].
b tce stands for a ton of standard coal equivalent. c China's Energy Development Strategy Action Plan for 2014–2020 released on November 19, 2014.

Table 3.
China's renewable energy development goals by 2050a.

6. China's renewable energy goals by 2050

China has turned out to be the world leader in renewable energy and made a great investment. According to [41] report, China promised US$286 billion for the development of renewable energy as well as US$376 billion for energy conservation projects in 2011–2015. According to [42] report, the total investment of the world in 2015 was US$286 billion with China contributing 102.9 billion in 2015 as the world's largest investor in the renewable energy development. China hosted over 25% of the world's non-hydro renewable capacity as at the end of 2015, being 63.1 and 117.0% higher than the United States and Germany, correspondingly [43]. In 2006, China enacted the Renewable Energy Law and Pricing Law that implemented tax reduction, financial support, subsidy policies, and measures for renewable energy applications [37].

Summary of the renewable energy targets by the Research Committee of China has been tabulated in **Table 3** (G1–G3) considering three cases comprising 2020, 2030, and 2050. The high uptake scenario would be a challenge to achieve. Nevertheless, there is a possibility of achieving it in realistic for China to achieve 45% of their energy from renewables to increase the county's total energy consumption of 5.8 billion tce by 2050 (G4 in **Table 3**) [44].

7. Conclusions

This manuscript has discussed the various renewable resource potentials in China. It was revealed that hydroelectric power has the highest installed and development potential in the country with its development in most of the cities and provinces in the country. Other resources such as wind, solar, and biomass were also observed to have a great potential in the country which are yet to be developed. The various techniques to eliminate coal and fossil fuel use to overcome pollution and emission control by the government have also been discussed. This collaboration will continue to lead and guide the world toward further green development and make massive green supports for the welfare of people. The various Development Strategic Action Plan for and Instantaneous of renewable energy targets by the various Research Committees in the country have also been analyzed and suggested appropriate solutions for achievement.

Acknowledgements

The authors would like to express their gratitude to Mrs. Joyce Adu, Mrs. Joyce Oppong, Dr. Asare Bediako, Sam Adu-Kumi Jnr., and Majeed Usman Koranteg for their proof reading and correction. This work was supported by the Special Funds of the National Social Science fund of China, grant number 18VSJ038.

Conflict of interest

The authors have complied with ethical requirements: submission implies that, if accepted, it will not be published elsewhere in the same form, in English or in any other language, without the written consent of the publisher. The authors have no conflict of interest to declare.

Thank you for your consideration of the work.

Author details

Daniel Adu[1*], Ransford O. Darko[2*], Boamah Kofi Baah[3] and Agnes Abbey[4]

1 School of Management Science and Engineering, Jiangsu University, Zhenjiang, China

2 Department of Agricultural University, University of Cape Coast, Ghana

3 University of Professional Studies, Accra, Ghana

4 Department of Renewable Energy, Cape Coast Technical University, Cape Coast, Ghana

*Address all correspondence to: adudaniel39@yahoo.com and chiefrodark@yahoo.com

References

[1] Ouyang XL, Lin BQ. Levelized cost of electricity (LCOE) of renewable energies and required subsidies in China. Energy Policy. 2014;**70**:64e73

[2] Ministry of Finance. National Development and Reform Commission, National Energy Administration (MOF/NDRC/NEA), Tentative procedures for the management of collecting and using renewable energy development special fund. January 2012

[3] National Development and Reform Commission. Ministry of Finance (NDRC/MOF), implementation opinions on promoting the development of wind power generation industry. 2006.

[4] Ministry of Finance (MOF). Tentative procedures for the management of special fund on the industrialization of wind power generating equipment. 2008.

[5] Ministry of Finance. Ministry of Science and Technology, National Energy Administration (MOF/MOST/NEA), Notice on Implementing the Golden Sun Demonstration Project, July 2009

[6] Zhao ZY, Zuo J, Fan LL, Zillante G. Impacts of renewable energy regulations on the structure of power generation in China e a critical analysis. Renewable Energy. 2011;**36**(1):24e30

[7] Liu LQ, Liu CX, Wang JS. Deliberating on renewable and sustainable energy policies in China. Renewable and Sustainable Energy Reviews. 2013;**17**:191e198

[8] BP Statistical Review of World Energy. June 2017. Available from: https://www.bp.com/content/dam/bp/en/corporate/pdf/energy-economics/statistical-review-2017/bp-statistical-review-of-world-energy-2017-full-report.pdf. [Accessed: 20 November 2017]

[9] Li F. China's energy structure continues to optimize. Available from: http://www.chinadaily.com.cn/business/2017-07/22/content_30211808.htm Accessed 20 December 2017

[10] BP. BP Statistical Review of China. London: BP; 2016

[11] CEC Sector Development and Environment and Resources Beijing, PR China. Available from: web@cec.org.cn

[12] IEA – Report. 2015. Available from: www.iea.org. Retrieved 2017-09-23.

[13] China Energy Portal. Tracking China's transition to sustainable energy. February 27, 2017

[14] CEC. National Power Industry Statistical Bulletin Data List. Beijing, PR China: Planning and Development Department; 2016

[15] Zhang J, Adu D, et al. Review on China's renewable energy and future Projections. International Journal of Smart Grid and Clean Energy. 2018;7(3):361-366

[16] IHA. Key Trends in Hydropower. Chancery House, St Nicholas Way, Sutton, London: Central Office; 2017

[17] Thompson R. China: Clean and Renewable Energy Report to. 2010

[18] Yang JX. Hydropower: A Viable Solution for China's Energy Future? 2007. World Watch Institute. Available from: http://www.worldwatch.org/node/4908

[19] Dams in China. Wikipedia.org. Available from: http://en.wikipedia.org/wiki/Category:Dams_in_China

[20] Renewable Energy in China: Grid Connected Wind Power. National Renewable Energy Lab. Available from: http://www.nrel.gov/docs/fy04osti/35789.pdf

[21] Gang W. Wind matters: China's role in the future of wind (PDF). Goldwind. 28 October 2015. Retrieved 8 February 2016.

[22] China's new wind power capacity hits record high - Xinhua — English.news.cn. Available from: news.xinhuanet.com. Retrieved 2016-02-08

[23] Xinhua, China's wind power capacity continues to grow. Available from: http://www.globaltimes.cn/content/1030911.shtml/ [dead link accessed 24/06/2017]

[24] Daniel C. Climate Wire, China blows past the U.S. in wind power. Scientific American. Retrieved from: 2016-02-08

[25] Gow D. Wind power becomes Europe's fastest growing energy source. Tuesday February 3, 2009. Gardian. co.uk. Available from: http://www.guardian.co.uk/environment/2009/feb/03/wind-power-eu

[26] Wind Electricity Installed Capacity. International Energy Statistics. Energy Information Administration (EIA)

[27] Nicholas S, He J. How China became a climate leader. 2014. Available from: Eco Business http://www.eco-business.com/opinion/how-china-became-climate-leader/. Accessed 20/09/2017

[28] REN21. Renewables 2016 global status report. Renewable energy policy network for the 21st Century. 2016

[29] Blauvelt, E. Biomass in China. World Energy Discussion. Jan 14, 2008. Available from: http://worldenergydiscussion.blogspot.com/2008/01/biomass-in-china.html

[30] Nature Editorial. Trend watch renewable boom. Nature. 2016b;**534**:157

[31] Zhang D et al. Present situation and future prospect of renewable energy in China. Renewable and Sustainable Energy Reviews. 2017;**76**:865-871

[32] China is adding solar power at a record pace. Available from: Bloomberg.com. 19 July 2017. Retrieved 1 August 2017

[33] World Watch Institute. Solar energy booming in China. Available from: worldwatch@worldwatch.org 1400 16th St. NW, Ste. 430, Washington, DC 20036, access November 22, 2017

[34] Zhang D, Li W, Lin Y. Wave energy in China: Current status and perspectives. Renewable Energy. 2009;**34**:2089-2092

[35] Liu T, Xu G, Cai P, Tian L, Huang Q. Development forecast of renewable energy power generation in China and its influence on the GHG control strategy of the country. Renewable Energy. 2011;**36**:1284-1292

[36] NRDC White Paper – Improving China's existing renewable energy legal framework; lessons from the International and Domestic Experience, Natural Resources Defense Council. October 2010

[37] Jin Yang X, Hu H, et al. China's renewable energy goals by 2050. Environmental Development. 2016;**20**:83-90 ISSN 2211-4645

[38] IEA. World Energy Outlook 2017. Available from: https://www.iea.org/weo/china/. Accessed 20 December 2017

[39] Wang S, Yuan P, Li D, Jiao Y. An overview of ocean renewable energy in China. Renewable and Sustainable Energy Reviews. 2011;**15**:91-111

[40] Here's how much money China is throwing at renewable energy. Available from: http://fortune.com. Retrieve 10 November 2017

[41] IRENA. Renewable Energy Prospects. China: International Renewable Energy Agency; 2017

[42] Energy Research Institute. China 2050 high renewable energy penetration scenario and roadmap study National Development and Reform Commission. 2015

[43] Kumar A, Kumar K, Kaushik N, Sharma S, Mishra S. Renewable energy in India: Current status and future potentials. Renewable and Sustainable Energy Reviews. 2010;**14**:2434-2442

[44] Nicholas S, He J. How China became a climate leader. 2014. Available from: http://www.eco-business.com/opinion/how-china-became-climate-leader/. Eco Business